Lecture Notes in
Computer Science

T0223683

Lecture Notes in Computer Science

Vol. 352: J. Díaz, F. Orejas (Eds.), TAPSOFT '89. Volume 2. Proceedings, 1989. X, 389 pages. 1989.

Vol. 353: S. Hölldobler, Foundations of Equational Logic Programming. X, 250 pages. 1989. (Subseries LNAI).

Vol. 354: J.W. de Bakker, W.-P. de Roever, G. Rozenberg (Eds.), Linear Time, Branching Time and Partial Order in Logics and Models for Concurrency. VIII, 713 pages. 1989.

Vol. 355: N. Dershowitz (Ed.), Rewriting Techniques and Applications. Proceedings, 1989. VII, 579 pages. 1989.

Vol. 356: L. Huguet, A. Poli (Eds.), Applied Algebra, Algebraic Algorithms and Error-Correcting Codes. Proceedings, 1987. VI, 417 pages. 1989.

Vol. 357: T. Mora (Ed.), Applied Algebra, Algebraic Algorithms and Error-Correcting Codes. Proceedings, 1988. IX, 481 pages. 1989.

Vol. 358: P. Gianni (Ed.), Symbolic and Algebraic Computation. Proceedings, 1988. XI, 545 pages. 1989.

Vol. 359: D. Gawlick, M. Haynie, A. Reuter (Eds.), High Performance Transaction Systems. Proceedings, 1987. XII, 329 pages. 1989.

Vol. 360: H. Maurer (Ed.), Computer Assisted Learning – ICCAL '89. Proceedings, 1989. VII, 642 pages. 1989.

Vol. 361: S. Abiteboul, P.C. Fischer, H.-J. Schek (Eds.), Nested Relations and Complex Objects in Databases. VI, 323 pages. 1989.

Vol. 362: B. Lisper, Synthesizing Synchronous Systems by Static Scheduling in Space-Time. VI, 263 pages. 1989.

Vol. 363: A.R. Meyer, M.A. Taitslin (Eds.), Logic at Botik '89. Proceedings, 1989. X, 289 pages. 1989.

Vol. 364: J. Demetrovics, B. Thalheim (Eds.), MFDBS 89. Proceedings, 1989. VI, 428 pages. 1989.

Vol. 365: E. Odijk, M. Rem, J.-C. Syre (Eds.), PARLE '89. Parallel Architectures and Languages Europe. Volume I. Proceedings, 1989. XIII, 478 pages. 1989.

Vol. 366: E. Odijk, M. Rem, J.-C. Syre (Eds.), PARLE '89. Parallel Architectures and Languages Europe. Volume II. Proceedings, 1989. XIII, 442 pages. 1989.

Vol. 367: W. Litwin, H.-J. Schek (Eds.), Foundations of Data Organization and Algorithms. Proceedings, 1989. VIII, 531 pages. 1989.

Vol. 368: H. Boral, P. Faudemay (Eds.), IWDM '89, Database Machines. Proceedings, 1989. VI, 387 pages. 1989.

Vol. 369: D. Taubner, Finite Representations of CCS and TCSP Programs by Automata and Petri Nets. X. 168 pages. 1989.

Vol. 370: Ch. Meinel, Modified Branching Programs and Their Computational Power. VI, 132 pages. 1989.

Vol. 371: D. Hammer (Ed.), Compiler Compilers and High Speed Compilation. Proceedings, 1988. VI, 242 pages. 1989.

Vol. 372: G. Ausiello, M. Dezani-Ciancaglini, S. Ronchi Della Rocca (Eds.), Automata, Languages and Programming. Proceedings, 1989. XI, 788 pages. 1989.

Vol. 373: T. Theoharis, Algorithms for Parallel Polygon Rendering. VIII, 147 pages. 1989.

Vol. 374: K.A. Robbins, S. Robbins, The Cray X-MP/Model 24. VI, 165 pages. 1989.

Vol. 375: J.L.A. van de Snepscheut (Ed.), Mathematics of Program Construction. Proceedings, 1989. VI, 421 pages. 1989.

Vol. 376: N.E. Gibbs (Ed.), Software Engineering Education. Proceedings, 1989. VII, 312 pages. 1989.

Vol. 377: M. Gross, D. Perrin (Eds.), Electronic Dictionaries and Automata in Computational Linguistics. Proceedings, 1987. V, 110 pages. 1989.

Vol. 378: J.H. Davenport (Ed.), EUROCAL '87. Proceedings, 1987. VIII, 499 pages. 1989.

Vol. 379: A. Kreczmar, G. Mirkowska (Eds.), Mathematical Foundations of Computer Science 1989. Proceedings, 1989. VIII, 605 pages. 1989.

Vol. 380: J. Csirik, J. Demetrovics, F. Gécseg (Eds.), Fundamentals of Computation Theory. Proceedings, 1989. XI, 493 pages. 1989.

Vol. 381: J. Dassow, J. Kelemen (Eds.), Machines, Languages, and Complexity. Proceedings, 1988. VI, 244 pages. 1989.

Vol. 382: F. Dehne, J.-R. Sack, N. Santoro (Eds.), Algorithms and Data Structures. WADS '89. Proceedings, 1989. IX, 592 pages. 1989.

Vol. 383: K. Furukawa, H. Tanaka, T. Fujisaki (Eds.), Logic Programming '88. Proceedings, 1988. VII, 251 pages. 1989 (Subseries LNAI).

Vol. 384: G.A. van Zee, J.G.G. van de Vorst (Eds.), Parallel Computing 1988. Proceedings, 1988. V, 135 pages. 1989.

Vol. 385: E. Börger, H. Kleine Büning, M.M. Richter (Eds.), CSL '88. Proceedings, 1988. VI, 399 pages. 1989.

Vol. 386: J.E. Pin (Ed.), Formal Properties of Finite Automata and Applications. Proceedings, 1988. VIII, 260 pages. 1989.

Vol. 387: C. Ghezzi, J.A. McDermid (Eds.), ESEC '89. 2nd European Software Engineering Conference. Proceedings, 1989. VI, 496 pages. 1989.

Vol. 388: G. Cohen, J. Wolfmann (Eds.), Coding Theory and Applications. Proceedings, 1988. IX, 329 pages. 1989.

Vol. 389: D.H. Pitt, D.E. Rydeheard, P. Dybjer, A.M. Pitts, A. Poigné (Eds.), Category Theory and Computer Science. Proceedings, 1989. VI, 365 pages. 1989.

Vol. 390: J.P. Martins, E.M. Morgado (Eds.), EPIA 89. Proceedings, 1989. XII, 400 pages. 1989 (Subseries LNAI).

Vol. 391: J.-D. Boissonnat, J.-P. Laumond (Eds.), Geometry and Robotics. Proceedings, 1988. VI, 413 pages. 1989.

Vol. 392: J.-C. Bermond, M. Raynal (Eds.), Distributed Algorithms. Proceedings, 1989. VI, 315 pages. 1989.

Vol. 393: H. Ehrig, H. Herrlich, H.-J. Kreowski, G. Preuß (Eds.), Categorical Methods in Computer Science. VI, 350 pages. 1989.

Vol. 394: M. Wirsing, J.A. Bergstra (Eds.), Algebraic Methods: Theory, Tools and Applications. VI, 558 pages. 1989.

Vol. 395: M. Schmidt-Schauß, Computational Aspects of an Order-Sorted Logic with Term Declarations. VIII, 171 pages. 1989 (Subseries LNAI).

Vol. 396: T.A. Berson, T. Beth (Eds.), Local Area Network Security. Proceedings, 1989. IX, 152 pages. 1989.

Vol. 397: K.P. Jantke (Ed.), Analogical and Inductive Inference. Proceedings, 1989. IX, 338 pages. 1989 (Subseries LNAI).

Vol. 398: B. Banieqbal, H. Barringer, A. Pnueli (Eds.), Temporal Logic in Specification. Proceedings, 1987. VI, 448 pages. 1989.

Vol. 399: V. Cantoni, R. Creutzburg, S. Levialdi, G. Wolf (Eds.), Recent Issues in Pattern Analysis and Recognition. VII, 400 pages. 1989.

Vol. 400: R. Klein, Concrete and Abstract Voronoi Diagrams. IV, 167 pages. 1989.

Vol. 401: H. Djidjev (Ed.), Optimal Algorithms. Proceedings, 1989. VI, 308 pages. 1989.

Vol. 402: T.P. Bagchi, V.K. Chaudhri, Interactive Relational Database Design. XI, 186 pages. 1989.

Vol. 403: S. Goldwasser (Ed.), Advances in Cryptology – CRYPTO '88. Proceedings, 1988. XI, 591 pages. 1990.

Vol. 404: J. Beer, Concepts, Design, and Performance Analysis of a Parallel Prolog Machine. VI, 128 pages. 1989.

Vol. 405: C.E. Veni Madhavan (Ed.), Foundations of Software Technology and Theoretical Computer Science. Proceedings, 1989. VIII, 339 pages. 1989.

Vol. 406: C.J. Barter, M.J. Brooks (Eds.), AI '88. Proceedings, 1988. VIII, 463 pages. 1990 (Subseries LNAI).

Vol. 407: J. Sifakis (Ed.), Automatic Verification Methods for Finite State Systems. Proceedings, 1989. VII, 382 pages. 1990.

Lecture Notes in Computer Science

Edited by G. Goos and J. Hartmanis

471

Beng Chin Ooi

Efficient Query Processing in Geographic Information Systems

 Springer-Verlag

Berlin Heidelberg New York London
Paris Tokyo Hong Kong Barcelona

Author

Beng Chin Ooi
Computer Science Department, Monash University
Clayton, Victoria, Australia 3168

Current address:
Institute of Systems Science, National University of Singapore
Heng Mui Keng Terrace, Kent Ridge, Singapore 0511

CR Subject Classification (1987): E.1, H.2.3, H.3.1–3

ISBN 3-540-53474-1 Springer-Verlag Berlin Heidelberg New York
ISBN 0-387-53474-1 Springer-Verlag New York Berlin Heidelberg

Printing and binding: Druckhaus Beltz, Hemsbach/Bergstr.
2145/3140-543210 – Printed on acid-free paper

Preface

Geographic Information Systems (GISs) are database systems that allow the manipulation, storage, retrieval, and analysis of geographic data and the display of data in the form of maps. In such a system, the database describes a collection of geographic entities over a two-dimensional map. The information describes entities that have a physical location and extent in some spatial region of interest, and queries involve identification of these entities based on their aspatial and spatial attributes, and the spatial relationships between entities. Conventional query languages must be augmented to contain spatial operators [ChF81] and additional spatial data structures introduced to support efficient geometric operations. In addition, new optimization strategies are required to process the resultant queries efficiently.

One of the important characteristics of relational DBMSs is the optimizer which automatically translates a query expressed in a nonprocedural language into an optimal sequence of access operations to evaluate the query. In such a system, the user need not know the physical structure of the database. Instead the optimizer estimates the cost of possible alternatives for processing the transaction based on the given physical structure of the database and computes the minimum cost sequence. However, existing DBMSs are not designed to support the hybrid queries that involve the selection of data based on spatial relationships.

Efficient indexing structures provide the query optimizer with the means to construct an efficient query execution plan. A new indexing structure called the *spatial kd-tree* (skd-tree) is proposed in this thesis. The new structure supports two types of proximity search, namely the *containment* search and the *intersection* search in order to facilitate the evaluation of queries involving spatial (geometric) operators.

The architecture of the GIS proposed in this thesis utilizes a conventional relational DBMS. The conventional DBMS provides efficient storage and retrieval of aspatial data. A spatial processor is included to process the spatial components of the queries. A global optimizer is proposed to sequence the evaluation of spatial and aspatial components of queries. This involves the decomposition of queries into aspatial and spatial subqueries, the ordering of these subqueries, and the merging of partial results. Among the subquery sequences, the best can be selected. The effectiveness of the method is demonstrated.

This book is the revised and extended version of a Ph.D. dissertation [Ooi88] submitted to the Computer Science Department, Monash University, Victoria, Australia. I am indebted to a number of people who have assisted me in one way or another to materialize this thesis. In particular, I would like to express my appreciation to my two supervisors, Dr. Ken J. McDonell of Pyramid Australia and Dr. Ron Sacks-Davis of Royal Melbourne Institute of Technology, for their patience, invaluable supervision and guidance throughout the development of this project. I would also like to thank Ron and Ken for reading and checking the correctness of the thesis, and for improving its presentation.

My sincere thanks are due to my departmental supervisor, Dr. Binh Pham, for reading and providing comments on the initial draft of the thesis, Dr. Bala Srinivasan, for useful discussions, Mr. Anthony Maeder for his advice, Dr. David J. Abel of CSIRO in Canberra, Australia for discussions and help, the programmers, Mr. David Hook and Mr. Vincent Trinh, who were supported through a collaborative project funded by CSIRO, for their feedback on the skd-tree algorithms, Dr. Hans Woessner and Mr. J. Andrew Ross of Springer-Verlag for their editorial help, and anonymous referees for their useful suggestions.

To Clement Chan, Nick Ch'ng, Guat Khunn Goh, Mark Goodwin, A.S.M. Sajeev and Thompson Thong, I offer thanks for discussions. I would also like to thank Guat Khunn Goh, Shirley Ho and A.S.M. Sajeev, for reading the initial drafts of the thesis.

Thanks to Monash University for financial support in terms of a Monash Graduate Scholarship, CSIRO, Canberra Australia for funding the project, and Institute of Systems Science, National University of Singapore for supporting me in preparing this manuscript.

To my wife, my mother, brothers and sisters, Po Sim and Kian Fah in particular, and other family members, I give thanks for their love, support and encouragement. This book is dedicated to my late father, Mr. Ooi Arm, whose love will always be remembered.

Singapore, August 1990   Beng Chin Ooi

Table of Contents

Chapter 1

Introduction

Spatial database research requires the integration of ideas and techniques from many areas within computer science such as computer graphics, image processing, artificial intelligence, and database methodology as well as from the traditional area of photogrammetry. (D.M. McKeown, Jr [McK84, p19])

1.1 Geographic Information Systems

In diverse applications, information systems are being used increasingly as a means to manage, retrieve and store the large quantities of data which are difficult to handle manually. Spatial data handling is no exception; systems for these applications are known as Geographic Information Systems (GISs), and these systems exhibit a range of requirements and techniques known collectively as geographic information processing [CCK81, NaW79]. The applications include cartography, demography, town/urban planning, navigation and natural resource management.

In what follows, we limit our discussion to a two-dimensional space even though we develop our work for a general multi-dimensional space.

Geographic information systems are database systems that allow the manipulation, storage, retrieval and analysis of geographic data as well as the display of data in the form of maps. In such a system, the database describes a collection of geographic objects over a two-dimensional map. Each geographic object can be generally classified as belonging to a particular entity class such as city, road, lake, etc. The objects are described by their associated aspatial (alphanumeric) attributes (e.g. population, name, usage, etc.) as well as their spatial attributes (e.g. location). Furthermore, they may be grouped into three generic spatial object classes namely, **point**, **line** and **region**. The classification of an object into the above three classes is closely related to its *extent*. The extent, which is the area of interest, may vary according to the application. For instance, while a city may be a point in a nationwide or worldwide geographic system, it may be a region in a statewide geographic system. On a given map, the geographic objects may *intersect*, may be *adjacent* to others, and may *contain* other objects. These relationships are termed *spatial relationships*, which are object oriented rather than attribute based. A spatial relationship such as *intersect* must be used to answer queries of the form "Which roads intersect with Wellington Road ?".

1.2 Motivation

In a GIS, the information describes entities that have a physical location and extent in some spatial region of interest and queries involve identification of these entities based on their aspatial and spatial attributes, and relationships (both spatial and aspatial) between entities. Conventional query languages **must** be *augmented* [ChF81] and additional spatial data structures introduced to support efficient geometric operations.

It is highly desirable that a GIS provides a storage and information architecture that integrates both the spatial and aspatial components of the database. An integrated system would allow a retrieval that is based upon spatial and aspatial characteristics. Hence, the interface language must be powerful enough to express a query involving both spatial and aspatial components.

In general, from the users' perspective, a GIS should have the following desirable characteristics:

(1) An external interface language that is capable of supporting:

  (a) queries that *select* objects of interest satisfying some criteria;
    Criteria may be:
    — aspatial (over any attribute)
    — spatial
    — an arbitrary combination of the above.

  (b) queries that may range over any stored information;

  (c) the display of selected objects and their attributes in graphical or textual representation.

(2) A windowing facility should be provided:

  (a) to view a particular area of interest;

  (b) to act as a top level filtering device to query only a small section of a global map space or to accentuate the scope of a query. This obviates the problem of having to specify a query window explicitly in the query formulation.

  The window definition is likely to be via a graphical input such as a mouse, a light-pen or a joystick.

(3) The user interface and query language should be insensitive to the spatial class of the objects; e.g. underlying geometric representation of objects as points, lines or regions should not affect query semantics. Neither the graphical and spatial data nor the scope of queries should be divided spatially into "map sheets". Were it deemed essential that the graphical data or spatial data to be fragmented, it should not impinge on the user view of the database.

(4) In some cases, two or more maps describing different entities but covering the same area are overlaid to generate a third unstored map that may be more sensible to its applications. As such, the users should be allowed to select which maps be displayed.

(5) The system should support a flexible interpretation of errors in spatial information arising from an inaccuracy in raw data or the use of an appropriate scale relative to the window size. For example, the coastline may be smoothed when a map of Australia is displayed, but would be shown in full detail if only a small part of it is displayed.

It is apparent from the above conditions that two separate processing units are required: a query processor and a graphics processor. While the query processor is responsible for the retrieval of spatial and aspatial data from the database, the graphical processor is responsible for the display and input of graphical data. With respect to the above five characteristics, Table 1.1 describes the responsible processor.

Table 1.1 Query tasking

GIS Characteristics	Responsible Processor(s)	
	Query Processor	Graphic Processor
1 (a)	√	
(b)	√	
(c)	√	√
2 (a)	√	√
(b)	√	√
3	√	
4	√	√
5		√

The Structured Query Language (SQL) [Cha74] has been proposed by the International Organization for Standardization [ISO86] as a basis for definition of a standard relational database language. Further, SQL-like interface facilities have been used in various commercial products (e.g. Database 2 [HaJ84], SQL/DS [Cod81], QMF [Sor84] and Oracle [Ora85]). For such reasons, SQL forms an ideal basis for extensions.

To handle spatial relationships, an extension of SQL, the GEOgraphic Query Language (GEOQL) [SMO87], is proposed to express queries involving spatial predicates. GEOQL supports both spatial and conventional aspatial predicates. As an example, the query "Find all roads that intersect with the roads which are adjacent to Monash University" may be expressed as follows:

```
SELECT   X.name
FROM     road X, road Y, region
WHERE    X intersects Y and
         Y is adjacent to region and
         region.name = 'Monash University'.
```

The response time for answering queries of the above nature is dependent on both the availability of auxiliary data structures (e.g. indexes) defined over both aspatial and spatial attributes and the degree to which query evaluation can be optimized. To this end, query optimization would have to integrate a great number of techniques including logical transformations of queries, optimization of access paths, and the storage and organization of data at the file system level.

Although there is a formidable body of literature on the use of indexing structures to facilitate query retrieval, most of these techniques are unsuitable for indexing spatial data, especially for objects that cover area. Hence, the first objective of this project is to investigate the data structures that are suitable for speeding up the evaluation of the spatial predicates. The other objective is to design an optimization strategy making use of appropriate indexing structures to evaluate GEOQL queries efficiently. Many similar extensions to SQL [AbS86, RoL84, RoL85] have been proposed, but the problem of optimization for these extended SQL variants has not been addressed in the literature. With this in mind, while designing an optimizer for GEOQL, every effort is taken to develop general, rather than language dependent, optimization strategies.

1.3 Indexing Structures

There is a great difference between aspatial data and spatial data in a GIS. Spatial relationships exist among the objects and it is possible to represent some of the frequently referenced relationships using specially constructed relations, parts of existing relations or by maintaining conventional links. However, although it is possible to pre-materialize some spatial relationships in this manner, it is not pragmatic to store all such relationships explicitly. Consequently, the dynamic evaluation of spatial relationships is necessary.

Conventional database management systems are able to process aspatial selection criteria efficiently, however, they are not well suited to the task of efficient evaluation of spatial relationships. Some sort of spatial indexing mechanism must be supported. Without a spatial index, a query such as *"Find all objects that are within a radius of 2 km of Mt. Buffalo"* may require a search of the whole database. This will be grossly inefficient compared to retrieving only objects in the vicinity of Mt. Buffalo; and a spatial indexing mechanism based on proximity can be used to prune the search space in this manner. Many structures have been proposed for spatial indexing, and a detailed survey is undertaken in this thesis. As a result of analyzing the strengths and weaknesses of existing structures, we propose a new data structure for spatial indexing, known as the **spatial kd-tree** (skd-tree). This new structure is based on the kd-tree [Ben75], which is an extension of binary trees.

1.4 Query Optimization

A special purpose GIS could be constructed to provide the full range of facilities outlined in Section 1.2. Such an implementation would require a huge development effort, and of necessity replicate many of the facilities already provided in existing general purpose systems (e.g. aspatial indexing, file organization, concurrency control, access control, etc.).

As mentioned previously, the existing conventional database management systems provide efficient storage and retrieval of aspatial data. Therefore, one approach to the design of a GIS is to extend such database systems to include spatial data and operators. Whilst the query language can readily be extended to provide the necessary additional facilities, efficient evaluation of queries with spatial or mixed selection predicates is a difficult task.

One of the well known strategies used in query optimization [WoY76] is query decomposition where a query is decomposed into a sequence of simpler subqueries, the semantics of this sequence of subqueries is identical to the semantics of the original query. We use this technique to decompose a query into subqueries such that each subquery involves either a totally spatial or a totally aspatial predicate. The decomposition allows us to use an existing SQL *backend* to efficiently evaluate aspatial subqueries. What is required now is a *spatial processor* using the supported spatial indexes to evaluate spatial subqueries.

When a query is decomposed to multiple subqueries, these multiple subqueries *must* be executed in a sequence that yields the correct answer. In some instances, a subquery *cannot* be processed before the completion of some other subqueries. This property is known as query dependency, and these dependencies can be represented by an acyclic dependency graph. On the other hand, certain subqueries may be processed independently, in which case the partial results must be subsequently merged. To handle these situations, rewriting rules are introduced to ensure the correctness of the final result.

Quite often, there are many ways to arrange the subqueries to produce a semantically acceptable result. The set of all possible subquery sequences may be too large to explore; without heuristics to reduce the possible search space, an optimization module which aims to improve the query evaluation efficiency may instead increase the overall processing costs. Thus a rule-based generator is developed to reduce the number of sequences generated by not considering sequences that are a priori suboptimal.

To further prune the search space, cost estimates are required. Using the information stored in a meta-database or a database catalog, the cost of each sequence can be estimated and the best sequence selected.

To execute the query, the subqueries of the selected sequence are executed sequentially. Of course, when permissible, subqueries that are not dependent on the completion of others may executed concurrently to improve the response time and to reduce page accesses.

1.5 Thesis Synopsis

In the next chapter, we review some GIS and query interfaces that have been proposed in the literature. Although many attempts to construct an efficient GIS have been made, only recently the focus has been to integrate the *access* of data (aspatial and spatial) [AbS86, RoL84, RoL85] rather than the *storage* of data. With this new approach, it is possible to facilitate the query processing by making use of spatial data structures that have been developed. In this thesis we propose an architecture that follows this model. Then we proceed to review optimization techniques, but we pay more attention to recently proposed methods.

The kd-tree proposed by Bentley [Ben75] is a generalization of the binary tree [Knu73] for indexing multi-attribute objects. Although it has been shown that the kd-tree [BER85c] is efficient for point data, the structure, as originally defined, is not capable of indexing multi-dimensional objects which often cover areas. Two techniques, namely object mapping [BaK86, Ros85] and object duplication [MHN84], have been proposed for extending kd-trees to overcome this problem. Object mapping techniques require k-dimensional objects to be stored as points in a $2k$-dimensional space, and object duplication techniques require the storing of an object in more than one location. Both methods introduce a number of problems. Chapter 3 presents a new structure, called the skd-tree, which avoids both object duplication and object mapping. The strength of the new structure is its efficient pruning during search. All associated operations are described in Chapter 3.

To show that the skd-tree is an efficient structure, a simulation study was undertaken to compare the skd-tree with the R-tree [Gut84], the kd-tree [BaK86] using the object mapping technique and the kd-tree [MHN84] using the object duplication technique. Experiments using both skewed and uniform data are used to show the relative efficiency of the skd-tree. The methodologies and results are presented in Chapter 4. Possible extensions of the skd-tree are also presented.

Some more recent proposals for optimization centre on the design of a general strategy suitable for an extensible DBMS [BaM86, GrD87, StR86]; this work which is at the experimental stage, aims to develop techniques flexible enough to accommodate different data models. Although we realize the importance of this new trend, we believe that another method of extending existing DBMS should be explored. In this respect, we investigate an easy and feasible way to extend the current existing DBMS for spatial data. A GEOQL query is broken into subqueries that are either totally spatial or aspatial so that totally aspatial subqueries can be executed by an existing SQL backend, and the spatial subqueries can be executed by the spatial processor. Once the subqueries are formed, they are executed in an order that minimizes the overall query cost. The strategy consists of the following four major steps:

(1) *Logical Transformation*: As part of the optimization, the query tree produced by parsing the initial query is rearranged so that the new representation is more amenable to efficient evaluation. As well as including conventional logical transformations (e.g. redundancy removal), a GEOQL query must be restructured so that the use of the spatial indexing is made possible during optimization.

(2) *Decomposition*: The parse tree produced by stage 1 is partitioned into subtrees which are either totally spatial or aspatial. Each subtree represents a subquery which must be executed by the query processor. The aspatial subqueries will be executed by the existing SQL backend and the spatial subqueries which cannot be processed by the SQL backend can be executed by the spatial processor.

(3) *Plans Formulation and Selection*: From the set of subqueries obtained in the previous step, different orders are considered. The best or cheapest is chosen among all subquery sequences.

(4) *Plan Execution*: From the chosen subquery sequence, SQL subqueries are passed to the SQL backend and spatial subqueries are passed to the spatial processor.

The same framework can also be used to extend existing DBMS for new applications like VLSI design and CAD/CAM without having much alteration. In Chapter 5, we present such an approach, and develop its extended query optimization strategy.

The query optimization model proposed in Chapter 5 was implemented on top of the Relational Test Bed [McD86] and is discussed in Chapter 6. The correctness and feasibility of the proposed optimization strategy are justified. Possible limitations and problems are also discussed in that chapter.

Chapter 7 concludes the thesis, highlighting the contributions and suggesting further work.

Chapter 2

Related Work and GEOQL

Designing a GIS inevitably includes the design of an end-user interface language, selecting efficient techniques for the storage, retrieval and manipulation of aspatial and spatial data, and display and storage of display data. The main intention of this chapter is to discuss the basic concepts, functions and characteristics of a GIS, and to present some of the important work related to the design of GIS databases that has appeared in the literature. We are more concerned with efficient query processing, consequently our survey is oriented towards query processing mechanisms in a GIS. With respect to this, we examine existing query languages, index structures and optimization strategies during query processing. Due to the breadth of these three topics, we do not attempt to cover all the proposals made in the literature. Further, when a relevant comprehensive survey of a particular area exists in the literature, we cover that area only briefly.

2.1 Geographic Information Systems

2.1.1 The Nature of Data

Before we proceed any further, it is essential to distinguish the types of data that would be stored in a GIS. By and large, the data can be classified as:

- *aspatial*: aspatial data are attribute-based data, which are alphanumeric (e.g. name and populations). These data are stored as relations.

- *spatial*: spatial data consist of coordinates and other spatial properties describing the spatial outline objects whose corresponding descriptions are stored as aspatial relations. These data are typically stored in an external file in a relational database system where records must be of fixed length. Using the object identitiers stored in a relation, the variable length description of spatial objects can be obtained.

- *graphical*: graphical data are mainly for display purposes, and are usually in the form of maps. It is within the map space that the spatial data is defined, and hence the graphical data is closely related to the spatial attributes of spatial data. The graphical data is expensive to capture, and hence it must be made available to as many applications as possible.

  While the graphical data is used mostly for the purpose of displaying the answer in the form of maps, it is the spatial and aspatial data that are manipulated by the query processor. Although the three are related, the organization and manipulation of each require different facilities. Existing GIS [CCK81] store spatial data as conventional relations in which a spatial object is represented by several tuples. Temporary spatial objects may need to be constructed for certain queries. Note that the storage of spatial, graphical and aspatial data in a single database inevitably slows down the access times.

Conventional data models are not particularly suitable for geographic applications because they do not efficiently support the type of operations that are required for geographical applications, and they are not suitable for the storage and manipulation of spatial data and graphical data.

In general, a GIS is different from a conventional DBMS in many ways [DaP85].

(1) The amount of data is usually very large (e.g. contour information) and the geographical data is expensive to capture.

(2) The data types consist of complex objects, like lines and regions.

(3) The spatial operators are often more complex than numeric operators.

(4) Spatial orderings of spatial objects are harder to define.

(5) The entities relationship is more complex in the sense that it must be computed or inferred.

2.1.2 On the Design of a GIS

Although it is widely recognized that conventional database systems cannot be used for data analysis involving images (pictures or maps), the techniques used in these conventional systems form a fundamental basis for designing a geographic information system. From a database point of view, a GIS must support integrated access; i.e. the interface query language must allow both spatial predicates and aspatial predicates, enabling users to retrieve data that satisfy either/both spatial or/and aspatial criteria. A spatial search is based on the proximity of objects of interest, while an attribute search is solely based on aspatial data (e.g. a predicate over the attributes name and population describing a city). Essentially, the answer to a query should be presented in the form of maps and/or relations. The correspondence between spatial objects on the map and the aspatial attributes describing those objects must be maintained to give the answers in the alternative forms. Although the access to spatial and aspatial data is integrated within a single query interface, the storage and processing of the two types of information are separate issues.

Two obvious approaches to the design of a GIS are given below.

(1) Design and implement a new special purpose system; this new system would support a special purpose query language and access mechanisms, extended data types and a special graphical processor to accommodate spatial and aspatial data storage and retrievals.

(2) Extend an existing system to accommodate management of spatial data. This approach would include the extension of a conventional DBMS to provide access to both spatial and aspatial data, and to communicate with a graphics subsystem responsible for all graphical input and output.

The first approach may provide a better overall performance due to the fact that it is designed to suit a specific application. However, it requires a comparatively larger implementation effort. Besides, constructing a monolithic application-specific system is against the trend of designing an extensible DBMS [BaM86, Fre87b, StR87], a DBMS that supports any user-defined data model without substantial implementation effort and facilitates a more complete data integration. With the second approach, we can make full use of a well-developed and technologically mature aspatial DBMS in order to reduce the implementation effort [AbS87, GaR88, OSM89, RFS88, SMO87].

2.1.3 Existing Systems

Many of the earlier systems [CCK81, NaW79] are oriented towards image processing. A database that stores images and descriptive data provides both image and non-image processing. Existing systems differ widely with respect to the operations and manipulation languages they support and are typically designed for specific applications. Broadly, the digital structures used by the systems can be classified into two classes, namely topological (linked organization) and grid structures (cellular organization). While topological structures define an ordered set of points and line segments, grid structures subdivide the area of interest into a rectangular grid. There are two major deficiencies with such data representations; first, the system implementer must view spatial entities in terms of a low level representation such as coordinate strings or grids. Second, query processing basically employs the technique of exhaustive search because entities cannot be retrieved based on proximity. Many researchers in this area [AbS84, AbS87, MHN84, SMO87, SRF87] have advocated that provision of a high level view of spatial entities is essential. In addition, more advanced techniques of query processing are required in order to provide efficient retrieval; these techniques operate on high level entities and exploit proximity relationships.

In what follows, we give a brief review of the existing systems. As the features of a database system consist of aspects like data modeling, optimization strategies, and the external language, it is senseless to solely discuss a system without touching other aspects. So the discussion of some systems are presented together with their external interface languages.

Over recent years, the relational database approach has become very widely adopted. Several systems that are built on top of existing relational DBMSs are GEO-QUEL [BeS77], AQL (A Query Language) [ABD80], GRAIN [Cha77, CLW80] and ADM (Aggregate Data Manager) [TII80]. AQL is a relational database system for business applications that has been extended for geographic applications. The language AQL is based on the high-level language called APL. In AQL, maps are divided into cells. Each cell is represented by a tuple that describes its boundaries and its identification. The concept of a hierarchical organization of entities (e.g. city is part of state, and state is part of country) is used and represented in the form of level in a map. GEO-QUEL [BeS77] is a special purpose geographic information system, one of the earlier systems built on top of an existing DBMS, constructed as a front-end to a well known system called INGRES [SWK76]. All maps in a database are treated as relations. Although the implementation effort is minimal, the system provides very few spatial operations. Another system that was built on top of another well known system called System R [ABC76] is ADM, an interactive image-oriented database system developed by IBM in Japan. Each image is defined as a data type and therefore treated as a data element in a relational database. With such an approach, the external relational interface language SEQUEL can be used to retrieve images just like conventional aspatial data. In the GRAIN system [CLW80, LiC79], the distinction is made between aspatial and spatial pictures. The aspatial data are stored in a relational database and their corresponding spatial graphical representation are stored separately in an "image store". The approach is different from other earlier attempts [BeS77] in the sense that GRAIN stores the spatial and aspatial separately to allow faster retrieval of spatial data. The GRAIN query language has natural language format, and its query has to be translated into a picture algebra query. The picture algebra is an augmented relational algebra to include spatial operators like *through*, *east*, *west* and *area*.

The GADS (Geo-data Analysis and Display System) [MaC80] was developed by IBM, San Jose, USA, and is also based on the relational data model. It was designed specifically for geographic applications. The database consists of a set of source files from which the data may be extracted into a tabular database. Over the extracted data the user may query and access using the decision support system. The query formulation is in the form of APL functions. The GDB (Geographical Data Base) system [Bur80] was developed by IBM in Mexico and supports an external interface language that is based on APL. Spatial and aspatial information are stored separately and are referred to as *positional* and *non-positional* data respectively. The database is viewed as a sequence of planes and each plane corresponds to an entity. Views may be defined over any subregions and any combination of planes. The image processors convert imaged based features to map-coordinate maps. MAPS (Map Assisted Photo Interpretation System) [McK84, McK83] is a large integrated database system, being developed in Carnegie-Mellon University, containing high resolution aerial photographs, digitized maps and other cartographic products. The system consists of several components; CONCEPTMAP is the database component of MAPS, in which the files contain the basic spatial information about maps, and the spatial data is hierarchically organized. In conjunction with other information stored in other components, MAPS supports a number of spatial query operations on data.

In summary, many of the earlier systems are very much image oriented and they have very limited capabilities for retrieving data based on spatial relationships. As expected, some of these special purpose systems have more facilities than extended systems [DaP85]. However, they tend to be expensive and a recent trend has been to develop extensible DBMSs.

A recent system built on top of ORACLE [Ora85] is the SIRO-DBMS [AbS87]. The DBMS is treated as a kernel and a set of procedures providing data manipulation tools comprise the shell of the kernel. Making use of an existing DBMS provides a straightforward implementation of the external interface language; the language SESQL (Spatial Extended SQL) is an extended SQL [Cha74]. The spatial description of an object is stored in the form of abstract data types [SRG83], known as pseudo-attributes. A spatial indexing structure [AbS84] is supported to facilitate retrieval based on spatial relationships.

A similar approach was adopted in [RFS88] and [GaR88]. Two commercial products that adopt such approach are ARC/INFO [ARC90] and Generation 5 [Gen90] systems. In ARC/INFO, the basic unit of data is the coverage. A coverage defines locational thematic attributes for map features in a given area. In the relational database of ARC/INFO, a coverage is defined by a set of relations. The relations define the geometric, topological, and attribute values of the various features in a coverage. Generation 5 system uses AutoCAD, a graphical user interface package, as the front end. The system supports an extended SQL and interfaces with a relational database system for aspatial data. In both systems, spatial and aspatial indexes are supported.

The trend towards extending existing DBMSs suggests that this approach may be feasible and becomes more popular than special-purpose systems. However, such an approach may not be efficient, and it is the main motivation for us to address the design based on such approach.

One of the proposed extensible DBMS products, the PROBE database system [OrM88] is object-oriented. There are two basic constructs in the PROBE data model:

entities and functions. A function is generically defined as a relationship between collections of entities and scalar values. The basic objects in a spatial data model in PROBE are point sets. In fact all objects are represented as linear objects [Ore86], a mapping of k-dimensional space to 1-dimensional space. Since spatial indexes, such as R-trees [Gut84], are not used, most geometric operations are of a nested-loop nature. The function of the "geometry-filter" in PROBE is to optimize the nested-loops in geometric algorithms. A prototype PROBE system has been completed and it is currently under experimentation [OrM88].

2.2 Query Languages

A database query language is an integral part of a DBMS, enabling users to interact with the system. A query language must be simple to learn and to use in order to simplify the task of query formulation, and yet it must be powerful enough to be useful. This is also true for geographic databases, since, although many users may be experts in their own areas (e.g. cartographers), they may not be skilled programmers. For such users, a geographic information system is just another tool with which to work. Hence ease of use is an important issue in designing an interface language to a geographic information system.

Although existing languages like SQL [Cha74, CAE76], QBE [Zlo76] and QUEL [HSW75] have gained recognition as powerful and useful languages, they have several weaknesses [Dat84b, Goh86, GoM87]. This highlights the difficulty in designing a new external interface language for a GIS. There are a number of advantages to extending an existing language. By extending a language, it is possible to make use of existing front-ends and query optimizers. In this section, we briefly discuss several external interface languages that are based on three of the major relational query languages, namely QBE, QUEL and SQL. We also briefly examine the proposed systems that support the external interfaces. It is assumed that readers are familiar with QBE [Zlo76], QUEL [HSW75] and SQL [Cha74].

2.2.1 QUEL Extensions

QUEL [HSW75] was developed at Berkeley, California as the external interface language to the INGRES database management system. QUEL is a tuple oriented language based on relational calculus and is modeled after the data-sublanguage ALPHA [Cod71b].

One of the earliest attempts to make use of an existing DBMS for geographical data processing was proposed by Berman and Stonebraker [BeS77]. The proposed special purpose language GEO-QUEL is constructed on top of the host system of QUEL, namely

INGRES [HSW75]. A map is stored in the INGRES as a relation of the following format:

R(X1, Y1, X2, Y2, PLZTYPE, other information)

where X1, Y1, X2 and Y2 are the coordinates of a point or a line, PLZTYPE is the type of the tuple indicating whether it is a line segment or a point, and the other attributes are for displaying purposes. With such a representation, all lines or regions have to be decomposed into a sequence of line segments. Maps can be manipulated either by special GEO-QUEL commands or by using the query language QUEL. Since the GEO-QUEL does not support geographic data types, it does not not allow retrievals that are based on spatial relationships, unless the relationships are precomputed and stored explicitly.

Some recent extensions to QUEL have been proposed for engineering applications, but they are however suitable for geographic applications. The notion abstract data type (ADT) used in a database context in [RoS79], is implemented in an extension of QUEL [SRG83, StG84, Sto86] for engineering applications. A column may be an ADT, which may be a character string in its internal representation that is materialized into its external representation. As an example, an ADT box may be represented by a string of integers, say "1,1,4,3", corresponding to two corner coordinates, (1,1) and (4,3), of a box. Spatial operations such as box intersection and containment are available as operators for ADT fields. The following query illustrates an example of QUEL with ADT:

```
range of b is boxes
retrieve (b.box-desc)
where b.box-desc * "0,1,3,4".
```

where box-desc is an ADT and * is the intersection operator. In addition to such a data type, new operators may be defined by users; these new operators must be written in C. A major problem with ADT is the expensive conversions from internal to external representation, and vice versa. In [SAH84], QUEL as data type is proposed, in which an attribute of a relation may be of type QUEL, i.e. an attribute may consist of a QUEL query that retrieves tuples from other relations. The intent of the extension is to provide a general referencing mechanism. For geographic applications, an object at a higher hierarchical [TKN80] level may have attributes of type QUEL which retrieves tuples from relations at lower hierarchical levels. For example, a relation *polygon* may have an attribute of type QUEL which retrieves tuples that describe line segments from the relation *line*. However, QUEL as a datatype is hard to update [KeW87]. The latest extension of QUEL is POSTQUEL [StR87], which is the interface language for the extensible DBMS called POSTGRES.

Prior to the proposal of QUEL with ADT, a similar extension to QUEL is proposed by Zaniolo [Zan83]. GEM supports new data types which correspond to pointers to tuples and sets of values. As pointed out in [KeW87], GEM has the same expressive power as "QUEL as a data type".

2.2.2 QBE Extensions

QBE, based on Codd's relational model [Cod70], was first developed by Zloof at IBM's research laboratories [Zlo76]. It is a two-dimensional language in which users are presented with a tabular representation of the underlying base relations of the database. To formulate a query, the user would just have to provide information in the appropriate rows and columns of the skeleton tables which are in the same format as the actual base tables of the database. The information filled in is just an example of a possible solution which consists of two elements, namely *constant elements* and *example elements*. The syntactic distinction between these elements is that an example element is underlined (or prefixed with an underscore) whereas a constant element is not.

Query-By-Pictorial-Example (QBPE), based on QBE, was proposed by Chang and Fu [ChF80a, ChF80b, ChF81] for a relational database system called IMAID. For each entity, two relations are used to store aspatial and spatial information. An aspatial relation stores the aspatial attributes, object identities and frame number. Corresponding to each tuple of an aspatial relation is a set of tuples that describes the spatial characteristics of objects. For example, for a region object with identity *id*, there will be multiple tuples with the same *id*, each describing a line segment, in the corresponding spatial relation. As with QBE, formulating a query essentially involves the filling of the skeleton tables by example elements and constant elements. However, apart from the base tables, intermediate tables are required to store partial results. Further, linear formulation is incorporated into the query formulation for spatial operations which are defined over intermediate tables.

Another extension of QBE was proposed in [BaB81] for a system called GEOBASE. A conceptual schema of a database is a hierarchy of lower level schemas. Syntactic and semantic constraints are imposed on geometric data types and aspatial data types. These constraints determine the spatial relationships among data (e.g. a country schema contains other entities like roads and states, which may in turn contain points), and inheritance of aspatial attributes (e.g. a governing party of a state may not be a governing party of a country, which is therefore not inheritable). For a query involving a schema and a subschema, a hierarchy of skeleton tables is used. Additional spatial functions are introduced, which may be used in the same way as an aggregate function, defining the spatial selection criteria in the condition box.

Although both QBE extensions allows the formulation of queries involving spatial selection criteria, both have stored spatial and aspatial data as relations. None of these proposals have addressed the problem of spatial indexing and query optimization.

A QBE-like query language PICQUERY [JoC88] was used as an external interface language in a pictorial database system called PICDMS (pictorial database management system). PICDMS [CCK81, CCK84] uses the gridded data representation scheme to store data; hence it is heavily oriented towards image processing. It is different from QPE [ChF80a] in that QPE uses topological structures for well delineated objects. The commands to manipulate pictures include the zooming of a picture, displaying and scanning parts of a picture, and superimposing a picture on the others. Associated with each command is a skeleton table, where fields (variables) are filled in. For example, superimposing objects from a picture with an object from another picture may be formulated as below:

Superimpose:

Picture 1	Object Name	Picture 2	Object Name
PIC1	A1	PIC2	A2

As stated in [JoC88], the open ended feature of the PICQUERY language is intended for extensions to suit a particular application.

2.2.3 SQL Extensions

The Structured English Query Language (SEQUEL), based on Specifying Queries as Relational Expressions (SQUARE) was developed by Chamberlin [Cha74] for the purpose of research. The major improvement of SEQUEL over SQUARE is the ease of use (obtained by replacing a positional format with a mathematical-like notation with a linear query language using English keywords). The new version of SEQUEL is SEQUEL2 [CAE76], which is better known as the Structured Query Language (SQL).

With strong support from IBM (DB2 [Dat84b]), acceptance in commercial circles (e.g. Database 2, [HaJ84] SQL/DS, [Cod81] QMF, [Sor84] and Oracle, [Ora85]), and the development of a draft ANSI standard [ISO86], SQL is emerging as the de facto interface language to relational database management systems. For these reasons, SQL forms an ideal base from which extensions may be made to support increased functionality [AbS86, RoL85, SMO87]. In fact, the two commercial GISs mentioned in Section 2.1.3, ARC/Info and Generation 5, support the interface to relational database systems in which the aspatial data is stored.

An earlier attempt to extend SQL for geographical applications is the MAPQUERY proposed by Frank [Fra82]. The system is a land information system and the CODASYL model is the underlying data model. Extra constructs were added to SQL to merely support facilities for graphical input and output, allowing windowing on the required area and representing object by signs. It does not support spatial operations.

Two SQL extensions [AbS86, RoL85] apart from our SQL extension [SMO87], have been proposed as an external interface language for GIS. We would describe the two extensions [AbS86, RoL85] briefly, and our extension [SMO87] in length since upon that language is this project based.

Pictorial SQL (PSQL) [RoL84, RoL85] has been proposed as an interface language for retrieving data from pictorial databases. The language has the following construct:

```
SELECT    list_of_attributes
FROM      list_of_relations
ON        list_of_picture
AT        area_specification
WHERE     qualification.
```

Two additional options are ON and AT, allowing users to select an area of a given picture. An area in the area-specification of the AT clause may be followed by one of the spatial operators, namely *covering*, *covered-by* and *overlapping*, and location specification or another query. The feature allows retrieval of information from two separate maps that are referring to the same geographical area.

Spatially Extended SQL (SESQL), an extended SQL, was proposed by Abel and Smith [AbS86] as an external interface language to a GIS called system SIRO-DBMS [AbS87]. One of the main operators introduced is *within*, which may have four different types of operands. In SIRO-DBMS, data types like *long* and *byte-string* are used for the description of spatial objects (e.g. a string of coordinate pairs to describe a region) in a relation. Further, a database is viewed as a hierarchical structure so that a large region may be partitioned into subregions.

2.2.4 GEOQL — An Augmented SQL

The GEOgraphic Query Language (GEOQL) is specifically designed for use in a GIS, and is an extension of SQL which provides spatial operators that may be used to express queries with both spatial and aspatial qualifications. For example, the query "Find all cities with a population over 5000 within a radius of 200 km of Mt. Buffalo" requires a spatial operator, WITHIN which limits the search space to a specified distance from a given object, which in this case is Mt. Buffalo.

This query can be expressed in GEOQL as follows:

```
SELECT   CITY.Name
FROM     CITY, MOUNTAIN
WHERE    MOUNTAIN.Name = 'Buffalo' and
         CITY.Population ≥ 5000 and
         CITY within 200 km of MOUNTAIN.
```

The operators supported by GEOQL are listed in Table 2.1. Note that some operators (e.g. *intersect*, *adjacent* and *joins*) are *commutative*, but others (e.g. *ends_at* and *contains*) are not commutative, i.e. Line A joins Line B ⇒ Line B joins Line A, but Line A ends_at Line B ⇏ Line B ends_at Line A.

The GEOQL language assumes a schema for the underlying relational database which guarantees that all aspatial information about objects of a particular entity class (e.g. city, lake, road, district) will be stored in a relation defined for that purpose and that each such relation holds information about objects of only one entity class. Consequently, a table (or variable) name can be used to identify the type of an operand (i.e. a geographic entity class) in a spatial expression.

The initial GEOQL [SMO87] does not provide a windowing facility which directly supports the retrieval of objects within a user specified query region. GEOQL is extended to provide such capabilities and a query region may be specified by its two corner coordinates or by two implicit coordinates which correspond to the current

Table 2.1 Spatial operators supported by GEOQL

GEOQL Operators
intersects
adjacent
joins
ends_at
contains
situated_at
within
closest
furthest

"window" (screen) or via a input device (e.g. light pen and mouse). If there is no predicate involving a window definition in a query and yet a "window" is specified via an input device, then the qualification in the query is conjuncted with the predicate *window contains geo_obj* for each geographic entity referenced in the query. Suppose while formulating the earlier query, a window is defined on a skeleton map. The resultant query is as follows:

```
SELECT   CITY.Name
FROM     CITY, MOUNTAIN
WHERE    MOUNTAIN.Name = 'Buffalo' and
         CITY.Population ≥ 5000 and
         CITY within 200 km of MOUNTAIN and
         window contains CITY and
         window contains MOUNTAIN.
```

With such an extension, zooming in on a portion of a region is possible. A window drawn on a displayed map as a result of a query retrieval will lead to the system displaying on a full screen size only part of the resultant map saved in a temporary storage.

The full syntax of GEOQL is given in Appendix A. In summary, SQL is extended to include *geo_pred* predicate as follows:

geo_pred	: geo_obj non_window_pred geo_term
	| window_term window_pred geo_term
	| window_term distance_pred geo_term
non_window_pred	: ENDS at_opt
	| JOINS
	| is_opt ADJACENT to_opt
	| is_opt SITUATED at_opt
	| is_opt WITHIN number of_opt
window_pred	: INTERSECTS
	| CONTAINS
distance_pred	: is_opt CLOSEST to_opt
	| is_opt FURTHEST from_opt
window_pred	: geo_term
	| WINDOW
	| window
geo_term	: geo_obj
	| LINE JOINING geo_obj AND geo_obj
	| geo_obj BOUNDED BY geo_obj AND geo_obj

geo_obj	: table_name
window_term	: (value_expr, value_expr, value_expr, value_expr)
is_opt	:
	| IS
at_opt	:
	| AT

The terms in capital letters are keywords.

A virtual entity may be specified as:

(a)   line joining geo_obj and geo_obj, or

(b)   geo_obj bounded by geo_obj and geo_obj.

In case (a) the two *geo_obj*s must be points, and in case (b), the first object must be a line or a region and the latter two *geo_obj*s must define points or lines that intersect the first *geo_obj* (or its boundary if it is a region). For example, a query to list all oil leases within a 5 km limit from the city Houston to Galveston is formulated as follows:

```
SELECT    *
FROM      lease, city CX, city CY
WHERE     lease within 5 of line joining CX to CY and
          CX.name = 'Houston' and
          CY.name = 'Galveston'.
```

The initial definition and implementation of GEOQL allows for only two-dimensional spatial objects and operators, and general extension of the spatial capabilities of GEOQL to three-dimensional data poses some semantic difficulties in the definition of the spatial operators. However, the basic GEOQL model can be extended to include three-dimensional data by adding the third coordinate value as an aspatial attribute of each object; the spatial objects would then remain defined over the XY-coordinate plane, but the Z coordinate could be queried using the standard SQL operators.

GEOQL supports three types of spatial objects, namely points, lines and regions. The data are not spatially divided into "map sheets" or other units, and queries may range over all the stored information. Restriction of the spatial area of interest may be provided by a windowing facility at the user interface, but this is not reflected in any internal partitioning of the stored data. While aspatial data are stored as relations, spatial data are stored in an external file. A unique identifier is attached to each geographic object and it is stored as an attribute in the relation. The identifier enables fast access to the variable length spatial description of an object.

A *line* consists of one or more line components. A *line component* is a connected set of straight-line segments. A line component consisting of n-1 segments can be represented by n XY co-ordinate pairs. A region is represented by one or more polygons. Each polygon has a positive or negative sense and defines a set of points consisting of the boundary and internal points of the polygon. Polygons with a *positive* sense describe a set of points to be *included* in the region, while those with a *negative* sense describe points to be *excluded*. Thus a region consisting of a number of (disjoint) islands can be represented by a set of positive polygons. In the example "a land area containing a lake", the land is the positive polygon and the enclosed lake is the negative polygon.

A sequence of polygons defining a region is subject to the following integrity constraints,

(a) each positive polygon must be *disjoint* with respect to the partial region defined by the preceding polygons, and

(b) each negative polygon must be *completely contained* within the partial region defined by the preceding polygons.

Apart from the augmented languages, many other special purpose languages have been proposed for the use in GIS; to name a few: [ChF81, CCK81, Har80, LiC79, Mei82, Mei85, NaW79, ShH78, TKN80]. There is great variation among these languages in how geographic entities are represented and manipulated.

2.3 The Need for Efficient Spatial Indexing Mechanisms

The spatial access method (the data structures and search algorithms to retrieve data by spatial keys) is a critical element of a spatial database management system as it is one of the major determinants of the overall performance of the system. (Abel, [Abe86, p164])

In a GIS or a CAD/CAM system, the database describes a collection of data objects over a particular multi-dimensional space. Efficient query processing relies upon auxiliary data structures used to support spatial indexing of these objects. The underlying data structure must provide efficient *geometric (spatial)* operations, such as locating the neighbors of an object and identifying objects in a defined query region.

It is widely recognized that conventional database systems are able to retrieve alphanumeric data efficiently, however, such database systems are typically unsuitable for efficient data access based upon spatial relationships. In a GIS, descriptive aspatial data is usually stored externally from the spatial data. With such structure, the aspatial data can be retrieved independently by conventional database management systems, and the spatial data can be used to support the spatial operations and operators that cannot be efficiently implemented within a conventional DBMS. Due to the cardinality of sets of data objects, it is highly inefficient to precompute and store spatial relationships among all data objects. Instead spatial relationships are materialized dynamically as required during query processing. In order to find spatial objects based on proximity, an indexing mechanism over the spatial location is considered essential. This section reviews the spatial indexing structures proposed in the literature. We review them in two ways; first, the structures are described, then the strengths and weaknesses are highlighted.

2.4 Point Structures

2.4.1 Introduction

Data structures of various types, such as B-trees [BaM72, Com79], ISAM indexes, hashing and binary trees [Knu73], are used as a means to permit efficient access, insertion and deletion of data. All these techniques are designed to index data based on the primary key. To use them to index data based on secondary keys, inverted indexes are required. However, this technique is not adequate for a database where range searching on secondary keys is one of the most common operations. For this type of application, multi-dimensional structures, such as grid-files [NHS84], multi-dimensional B-trees [Kri84, OuS81, ScO82], kd-trees [Ben75] and quad-trees [FiB74] have been proposed to index multi-attribute data. Such indexing structures are known as point indexing structures as they are designed to index data objects which are points in a multi-dimensional space.

Queries can be generally classified as exact-match or range queries. An exact-match query requires the point search to search for a particular point, whereas a range query requires a search for all points within specified ranges. Range searching has been a focus of a great deal of research in the past few years. A number of structures have been proposed for handling multi-dimensional data. This subsection describes two earlier structures. A comprehensive survey on earlier methods on range searching can be found in [BeF79].

2.4.2 Quad-Trees

2.4.2.1 Point Quad-Trees

The point quad-tree [FiB74] (see Figure 2.1) was proposed by Finkel and Bentley to represent data points in a multi-dimensional space; it can be viewed as a multi-dimensional generalization of the binary search tree [Knu73]. A node of a k-dimensional quad-tree stores a data point and may have 2^k sons, which in turn are the roots of the subtrees corresponding to the 2^k quadrants. Thus a 2-dimensional quad-tree node would have four sons, each representing a quadrant of four directions, namely, NW, NE, SW,

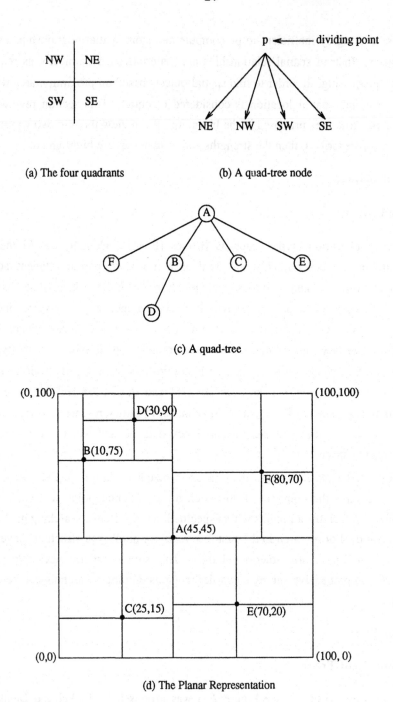

(a) The four quadrants

(b) A quad-tree node

(c) A quad-tree

(d) The Planar Representation

Figure 2.1 The structure of a 2-dimensional quad-tree

and SE. The four quadrants may span unequal areas and as data points determine the branching directions, no two points may have the same co-ordinates.

When performing an exact match search, a comparison is made at each node and the searching is stopped with a match; otherwise the subtree chosen for the next traversal is the quadrant in which the query point lies. Range queries are also well supported by point quad-trees, however it may be necessary to traverse multiple subnodes from any node that is visited during the search.

For insertion of a data point, the quad-tree is first searched to determine whether the data point is in the tree and if the data point is not found, it is inserted as a leaf node.

Deletion is not as easy as an insertion, because data points are stored in the internal nodes. When a node is deleted, the data points in the subtree must be reorganized. The method proposed is to reinsert all the data points in the subtree.

To make point quad-trees more dynamic, Overmars and Leeuwen [OvL82] introduced pseudo quad-trees which are non-homogeneous. Instead of data points, internal nodes are chosen as arbitrary points that suitably divide the subquadrants. Data points reside in the leaf nodes and this simplifies deletion.

2.4.2.2 Region Quad-Trees

The quad-tree representation of regions, proposed by Klinger [Kli71], has been the focus of research [AbS84, Mar82, Sam84, Sam86] in the fields of both image processing and database management. The region quad-tree (Figure 2.2) is a variant of the maximal block representation: a representation where the blocks must be disjoint and have a standard size which is a power of two. These characteristics allow a systematic way for representing homogeneous parts of an image and is known as regular decomposition. The repetitive pattern of partition of a quadrant into subquadrants enables quad-trees to represent images of any desired degree of resolution. It is not within our scope to give any further detail on quad-trees and a comprehensive survey on quad-trees can be found in [Sam84].

The properties of the quad-tree makes the data structure very suitable for image processing. Nevertheless, it can be also used to represent point data. Two suggested approaches are given below.

(1) Treat a data point as a black pixel in a region quad-tree [Sam84]. The representation can be thought of in terms of matrix representation where a data point represents a non-zero element in a square matrix. Such a representation is known the MatriX (MX) quad-tree. The organization of a matrix quad-tree is similar to that of a region quad-tree with the major distinction being the leaf nodes of a MX quad-tree are either black or empty (white) indicating the presence or absence of a data point

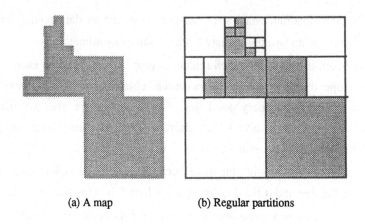

(a) A map   (b) Regular partitions

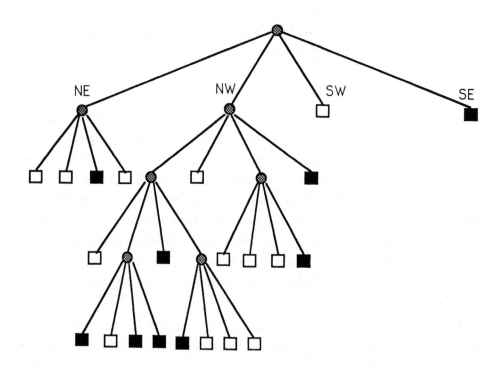

(c) A region quad-tree

Figure 2.2 Representation of a region by a region quad-tree

respectively. In a planar representation of such a quad-tree, a data point belongs to a quadrant if it is at the lower left corner of the square and each data point corresponds to a 1 by 1 square (see Figure 2.3). All nodes corresponding to data points appear at the same depth in the quad-tree.

(2) Treat each data point as a quadrant; that is, a quadrant must contain *at most* one point. Such a quad-tree is known as Point Region (PR) quad-tree [Ore82]. It is different from the MX quad-tree in that all data points may not necessarily reside on the same level, and a data point may be anywhere within the containing quadrant. Figure 2.4 illustrates a PR quad-tree structure.

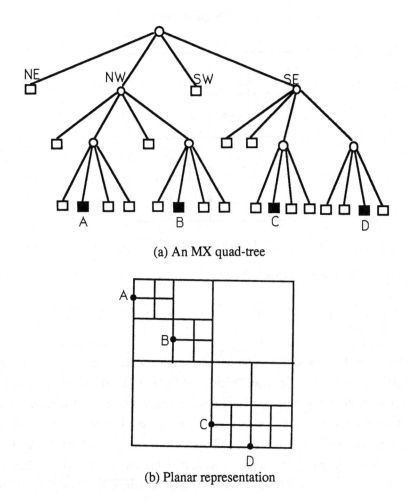

(a) An MX quad-tree

(b) Planar representation

Figure 2.3 The MX quad-tree structure

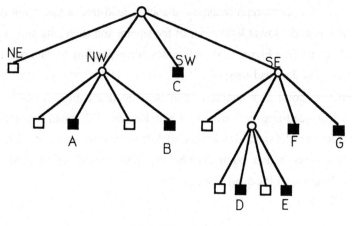

(a) A PR quad-tree

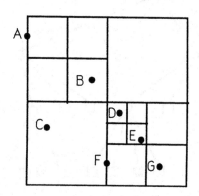

(b) Planar representation

Figure 2.4 The PR quad-tree structure

2.4.3 The kd-Trees

The kd-tree [Ben75], k-dimensional homogeneous binary search tree, was first proposed by Bentley. A node in the tree (Figures 2.5a and 2.5b) serves two purposes: representation of an actual data point and direction of a search. A discriminator whose value is between 1 and k inclusive, is used to indicate the key on which the branching decision depends. A node P has two children, a left son $LOSON(P)$ and a right son $HISON(P)$. If the discriminator value of node P is the jth attribute (key), then the jth attribute of any node in the $LOSON(P)$ is less than jth attribute of node P, and the jth attribute of any node in the $HISON(P)$ is greater than that of node P. This property enables the range along each dimension to be defined during a tree traversal and the ranges are smaller in the lower levels of the tree.

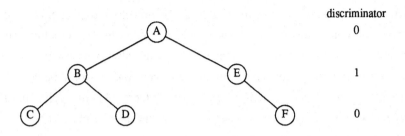

discriminator

0

1

0

Figure 2.5a The structure of a kd-tree

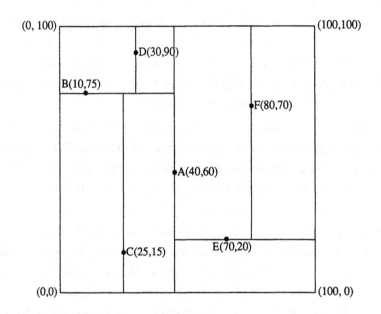

Figure 2.5b The planar representation

Complications arise when an internal node is deleted, although these are not as bad as in the case of the point quad-tree. When an internal node is deleted, say Q, one of the nodes in the subtree whose root is Q must be obtained to replace Q. Suppose i is the discriminator of node Q, then the replacement may be a node in the right subtree with the smallest ith attribute value in that subtree, or a node in the left subtree with the biggest ith attribute value. The replacement of a node may also cause successive replacements.

To reduce the cost of deletion, the non-homogeneous kd-tree [Ben79a] was proposed. The major difference is that when splitting an internal node, the non-homogeneous kd-trees selects an arbitrary hyperplane (a line for two dimensional space) to partition the data points into two groups having almost the same number of data points and all data points resides in the leaf nodes. Splitting (partitioning) on different dimensions cyclically as in the kd-tree [Ben75] is not always desirable. In cases where queries are mainly range queries on one of the dimensions, it would be desirable to split on that particular dimension more often than the others. Alternatively, splitting on the dimension with the longest interval yields data spaces that are more cubical. In general, in order to minimize the splitting process, the splitting dimension should be chosen such that data entries can be partitioned into two groups with approximately the same number of data entries.

The kd-tree has been the subject of intensive research [BaK86, BER85a, BER85b, BER85c, BeF79, Ben79b, ChF79, EaZ82, FBF77, LeW77, MHN84, Ore82, OvL82, Rob81, Ros85, ShB78, ShR85, etc.] over the past decade. Many variants have been proposed in the literature to improve the performance of the kd-tree with respect to issues such as clustering, searching, storage efficiency and balancing.

An indexing tree may sometimes be too large to be stored in main memory; it has to be paged into secondary storage (e.g. disks and tapes). B-trees [BaM72, Com79] are a very commonly used single attribute-based file organization for secondary storage. To improve the paging capability of the kd-tree, the K-D-B-Tree [Rob81] which is a combination of a kd-tree and a B-tree is proposed. The K-D-B-tree is height-balanced.

The K-D-B-tree consists of two basic structures: region pages and point pages (see Figure 2.6). While point pages contain object identifiers, region pages store the descriptions of subspaces which the data points are stored and pointers to descendant pages. In a non-homogeneous kd-tree [Ben79a], a space is associated with each node; a global space for the root node, an unpartitioned subspace for each leaf node. The subspaces stored in the region pages (e.g. s_{11}, s_{12} and s_{13}) are pairwise disjoint and together they span the rectangular subspace of the current region page (e.g. s_1), a subspace in the parent region page. For the following explanation, the elements in a region page (pointer and subspace description pairs) are referred to as entries.

During insertion of a new point into a full point page, a split will occur. The point page is split such that the two resultant point pages will contain the same number of data points. Note that a split of a point page requires an extra entry for new point page, this entry will be inserted into the parent region page. Therefore, the split of a point page may cause the parent region page to be split as well, which may further ripple all the way to the root; thus the tree is always perfectly balanced.

When a region page is split, the entries are partitioned into two groups. A hyperplane is used to split the space of a region page into two subspaces and this hyperplane may cut across the subspaces of some entries. Downward propagation of the split may occur as a consequence; the subspaces that intersect with the splitting

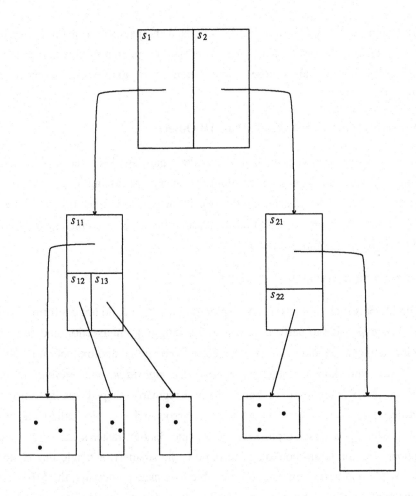

Figure 2.6 A K-D-B-tree structure

hyperplane must be split as well so that the new subspaces are totally contained in the resultant region pages. If the constraint of splitting a region page into two region pages containing about the same number of entries is not enforced, then downward propagation of split may be avoided.

The upward propagation of a split will not cause the underflow of pages but, the downward propagation is detrimental to storage efficiency because a page may contain less than the usual page threshold, typically half of the page capacity. To avoid unacceptably low storage utilization, local reorganization is performed. Two or more pages whose data spaces form a rectangular space can be merged and the resultant page is resplit if overflows. In Figure 2.6, the reorganization of s_{12} (s_{13}) only involves merging (and splitting) with s_{13} (s_{12}), whereas s_{11} requires to be reorganized with both s_{12} and s_{13}.

In summary, the K-D-B-tree has incorporated the pagination of the B-tree and the tree is height-balanced as a result. Nevertheless, the poorer storage efficiency is the trade-off. The structure, as proposed, cannot be used for non-zero sized objects such as lines or regions.

2.5 Access Structures for Extended Spatial Objects

In this section, we review most of the structures that have been recently proposed for indexing non-zero sized objects to improve query processing efficiency in spatial databases. As with other databases, the two basic issues that need to be addressed in selecting a data structure for a spatial database system are: *the efficient use of storage* and *the ease of information retrieval*.

2.5.1 Bounding Rectangles

In a GIS, objects such as roads, lakes, exploration leases, etc. do not conform to any fixed shape. It is expensive to perform any spatial testing (e.g. intersection and containment) on their exact location and extent and therefore, some initial approximation or filtering is used. Two well known methods, *region decomposition* and *minimum bounding rectangles* (MBR), are used to approximate irregularly shaped spatial objects. The region decomposition is best known with its associated structure called the quad-tree [Sam84], which is a tessellation of an object into disjoint raster squares of the desired resolution. The minimum bounding rectangle is the smallest rectangle that encloses the object. While region-decomposition is suitable for image processing, the MBR is widely used in proximity query processing. MBRs allow efficient proximity query processing by preserving the spatial identification and dynamically eliminating many potential intersection tests quickly. Two objects will not intersect if their MBRs do not intersect. This will reduce the cost since the test on the intersection of two polygons or a polygon and a sequence of line segments is expensive as compared to the test on the intersection of two rectangles. The k-dimensional MBR is defined as

$$(I_0, I_1, ..., I_{k-1})$$

where I_i is a closed bounded interval $[a, b]$ describing the extent of the spatial object along dimension i. It can be easily defined by a single dimensional array of size $2k$. Alternatively, the MBR of an object can be represented by its centroid and extensions on each of the k directions.

Objects extended diagonally may be badly approximated by MBRs, and false matches may result. That is, objects that are potentially eliminated during predicate testing are retrieved. Alternative proposals such as decomposition of regions into convex cells [Gun88] have been made to improve object approximation. The decomposition has its problem of having to store object identity in multiple locations in an index. In what follows, we use MBRs in our discussion. It should be noted that the approximation technique should not affect the design of an indexing structure.

2.5.2 Search Types

The main searching operations that have been addressed in the literature are:

- *Point Queries*: Given a point in the space, find all objects that contain it.
- *Region Queries*: Given a query region, find all objects that intersect it. The query region is usually rectangular, and is sometimes referred to as query rectangle or query window.

Searching for all regions that contain a given point is fundamentally the same as searching for all regions intersecting another region which has zero area. Hence we use the term *intersection search* to mean either type of search, and the point search as before to mean exact-match search of a point. A subclass of the intersection search is the *containment search*, which is a search for all regions that are strictly contained in a given query region. For all the indexing structures that follow, a containment search is treated in the same way as an intersection search, namely if the approximations to the regions intersect, containment is a possibility.

The term M is used to denote the capacity of a secondary page in terms of the number of entries, and m to denote the minimum threshold or fill factor.

2.5.3 Spatial Indexing Methods

Spatial access methods that follow can generally be classified into three groups [Ooi88, SeK88]. These methods are characterized by their techniques in extending a point indexing structure to a multi-dimensional spatial indexing structure. The performance of a structure is closely related to its extending technique(s).

(1) *Object Mapping (Transformation)*

Objects in a k-dimensional space are represented as points in a $2k$-dimensional space. For example a 2-dimensional rectangle described by (x_1, y_1, x_2, y_2), is represented as a point in 4-dimensional space, where each attribute is taken as from different dimension.

(2) *Object Duplication* or *Object Clipping*

The k-dimensional data space is partitioned into pairwise disjoint subspaces. With object duplication, an object is stored in all subspaces it intersects; that is, an object may be stored in multiple pages. A similar technique is based on object clipping. Instead of duplicating the identifier, an object is partitioned into several disjoint smaller objects so that each smaller object is totally included in a subspace.

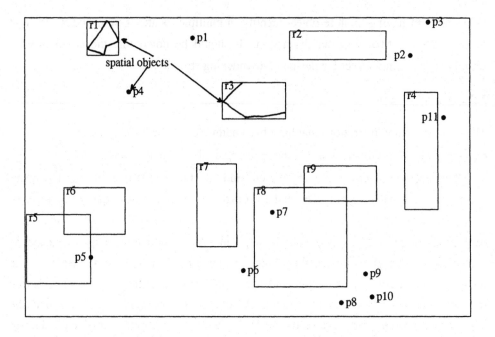

Figure 2.7 Spatial objects

(3) *Overlapping Bounding Rectangle*

Rather than supporting disjoint subspaces as in the object duplication scheme, the overlapping bounding rectangle scheme allows overlapping subspaces such that objects are totally included in one of the subspaces.

The example in Figure 2.7 will be used to illustrate the spatial coordinate representation of some of the spatial indexing structures that follow.

2.5.4 Corner Stitching

Corner stitching [Ous84] is a data structure that is specifically designed to support geometric operations in VLSI layout systems. The prime objective of the data structure is to link the rectangles according to their proximity. Rectangles are linked by pointers, called *corner stitches*, at two corners. Each rectangle is linked to four neighboring rectangles: one immediately above the top right corner, one on the right of the top right corner, one immediately below the bottom left corner and one on the left of the bottom left corner (illustrated in Figure 2.8). In the structure, not only the rectangular objects are stored but also the rectangular "space tiles" in between these object. Due to its linked

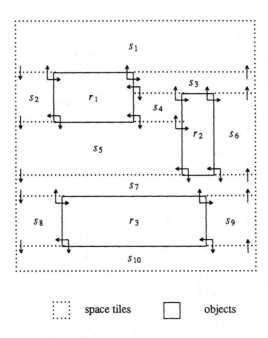

| ⁞ ⁞ | space tiles | ☐ | objects |

Figure 2.8 An example of tiles in corner-stitch data structure

structure, the structure is reasonably efficient for scanning the immediate neighborhood of a rectangle. However, it is expensive to locate rectangles of particular location where a linear search is essential. Since the structure maintains the 'space tiles' explicitly for which extra storage is required, searching is expensive. Further, overlapping rectangles are not allowed.

2.5.5 Cell Methods based on Dynamic Hashing

Both extendible hashing [FNP79] and linear hashing [KrS86, Lar78, SGR85] lend themselves to an adaptable cell method for organizing k-dimensional objects. The grid file [HiN83, NHS84] and the EXtendible CELL (EXCELL) method [Tam82a, Tam84b] are extensions of dynamic hashed organizations [FNP79] incorporating a multi-dimensional file organization for multi-attribute point data.

2.5.5.1 The Grid File

The *grid file* structure proposed in [HiN83, NHS84] consists of two basic structures: k linear *scales* and a k-dimensional *directory* (see Figure 2.9). The fundamental idea is to partition a k-dimensional space according to an orthogonal grid. The grid on a k-

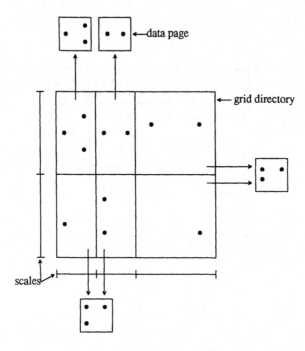

Figure 2.9 The grid file layout

dimensional data space is defined as scales which are represented by k one-dimensional arrays. Each boundary in a scale forms a $(k$-$1)$-dimensional hyperplane that cuts the data space into two subspaces. Boundaries form k-dimensional unpartitioned rectangular subspaces, which are represented by a k-dimensional array known as the *grid directory*. The correspondence between directory entries and grid cells (blocks) is one-to-one. Each grid cell in the grid directory contains the address of a secondary page, the *data page*, where the data objects that are within the grid cell are stored. As the structure does not have the constraint that each grid cell must at least contain m objects, a data page is allowed to store objects from several grid cells as long as the union of these grid cells together form a rectangular box, which is known as the *storage region*. These regions are pairwise disjoint, and together they span the data space. For most applications, the size of the directory dictates that it be stored on secondary storage, however, the scales are much smaller and may be cached in main memory.

Figure 2.10 illustrates a three dimensional grid object space. A 3-dimensional array, $dir(1..3, 1..3, 1..2)$, is required to store the grid entries, and the description of an entry may be obtained once the scales are known. For example, the description of the grid cell with $x_3 \times y_1 \times z_2$ subspace is stored in $dir(3, 1, 2)$.

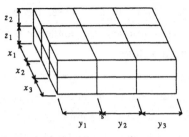

Figure 2.10 A three-dimensional grid object space

Like other tree structures, splitting and merging of data pages are required during insertion and deletion. Insertion of an object entails determining the correct grid cell and fetching the corresponding page followed by a simple insertion if the data page is not full. In the case where the page is full, a split is required. The split is simple if the storage region covers more than one grid cell; the grid cells are allocated to either the existing data page or a new page and and the data objects are distributed accordingly. However, if the page region covers only one grid cell then the grid has to be extended by a $(k-1)$-dimensional hyperplane that partitions the storage region into two subspaces. A new boundary is inserted into one of the k grid-scales to maintain the one-to-one correspondence between the grid and the grid directory, a $(k-1)$-dimensional cross-section is added into the grid directory. The two resulting storage regions are disjoint and, to each a new data page is attached. The objects stored in the overflowing page are distributed among the two new pages. Other grid cells that are partitioned by the new hyperplane are unaffected since both halves of the old grid cell will now be sharing the same data page.

When the joint occupancy of two adjacent storage regions drops below a certain threshold, the data pages are merged into one. Two different methods have been proposed for merging, the *neighbor* system and the *buddy* system. The neighbor system allows two data pages whose storage regions are adjacent to merge so long as the new storage region remains rectangular; this may lead to "dead space" where neighboring pages prevent any merging for a particular underpopulated page. A more restrictive merging policy like the buddy system is required to prevent the dead space. For the buddy system, two pages can be merged provided their storage regions can be obtained from any subsequent larger storage region using the splitting process.

The merging process will make the boundary along the two old pages redundant, if no storage regions are left that are adjacent to the boundary. Consider Figure 2.10, the boundary s becomes redundant if all storage regions that span over y_1 and y_2 intervals along dimension Y intersect with s. In such case, the redundant boundary may be removed from its scale and the one-to-one correspondence is maintained by removing the redundant entries from the grid directory.

A major problem of the grid file is the storage for the directory. The growth of the directory is exponentially proportional to the growth of the number of indexed objects. Most of these directory entries correspond to completely empty grid cells, i.e. they do not contain any data objects. A multi-level directory [Hin85, WhK85] has been suggested to alleviate the storage problem, however this approach does not support the property of "two disk accesses" [NHS84] for exact-match queries. The two disk accesses that are required for exact match queries are the accesses of the directory and the data page. The two disk assess property can only be obtained if the directory is stored as an array [NHS84, WhK85]. In this case, however, the size of the directory is doubled whenever a new boundary is introduced. Recently, the application of extendible hashing methods [Oto85] has been suggested for the grid file [Fre87] to improve the directory storage requirement without severe access cost penalties.

In [NiH85], the grid file is proposed as a means for spatial indexing of non-point objects. To index k-dimensional data objects, mapping from a k-dimensional space to a $2k$-dimensional space where objects exist as points is necessary. For example, a two-dimensional rectangle object is represented by a four attribute tuple (cx, cy, dx, dy), where (cx,cy) is the centroid of the object and (dx, dy) are the extensions of the object from the centroid.

One disadvantage of this scheme is that the higher dimensional space means it is harder to perform directory splitting [WhK85]. In addition, the retrieval of spatial non-zero sized objects based on spatial relationships can be fairly expensive. For an object (ocx, ocy, odx, ody) to intersect with the query region (qcx, qcy, qdx, qdy), the following condition must be satisfied:

$$ocx - odx \leq qcx + qdx \text{ and}$$
$$ocx + odx \geq qcx - qdx \text{ and}$$
$$ocy - ody \leq qcy + qdy \text{ and}$$
$$ocy + ody \geq qcy - qdy.$$

Consider Figure 2.11a, rectangle q is the query rectangle. The intersection search region on cx-dx hyperplane, the shaded region in Figure 2.11b, is obtained by the first two inequality equations of the above intersection condition. Note that the search region can be very large if the global space is large and the largest rectangle extension along the X axis is not defined. In Figure 2.11b, a knowledge of the upper bound, udx, for any rectangle extension along the X axis, reduces the search region to the enclosed shaded region. The same argument applies for the other co-ordinate. Objects that fall in both search regions satisfy the intersection condition.

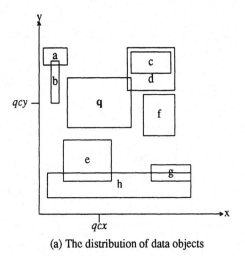

(a) The distribution of data objects

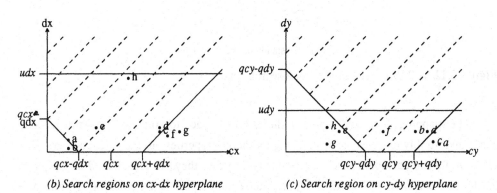

(b) Search regions on cx-dx hyperplane   (c) Search region on cy-dy hyperplane

Figure 2.11 *Intersection search region in the grid file*

It is easier to understand how the search may be performed from a 2-dimensional view point rather than from a 4-dimensional view point. Suppose the global 2-dimensional space of a database is defined and is represented as (0, 0) and (gx, gy). Further, the maximum extensions of objects are known, and let them be udx and udy respectively. Then the search region for the query defined by two corners, ($qx1$, $qy1$) and ($qx2$, $qy2$) is constrained by the following two equations:

$$\text{MAX}(\frac{qx1}{2}, qx1 - udx) < x < \text{MIN}(\frac{gx+qx2}{2}, qx2 + udx)$$

$$\text{MAX}(\frac{qy1}{2}, qy1 - udy) < y < \text{MIN}(\frac{gy+qy2}{2}, qy2 + udy)$$

where MAX and MIN functions respectively return the maximum and the minimum of their argument. Figure 2.12 illustrates the search space defined by the above condition. Objects whose centroids are in the shaded region will not intersect with the query region.

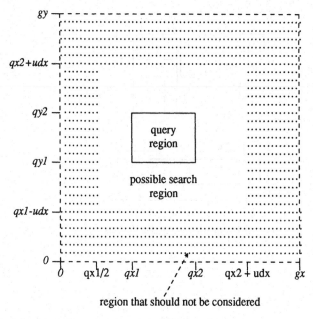

region that should not be considered

Figure 2.12 A 2-dimensional view of a search space

The mapping of regions from a k-dimensional space to points in a $2k$-dimensional space undesirably changes the spatial neighborhood properties. Regions that are spatially close in a k-dimensional space may be far apart when they are represented as points in $2k$-dimensional space. Consequently, the intersection search is not efficient. To improve the search performance of the grid file, a multi-layer grid file which avoids object mapping was proposed in [SiW88]. In such a structure, a map space may consist of several grid files that cover the same space. When a grid file is partitioned, all objects that are not cut by the partitioning hyperplane are distributed among the new two subspaces and objects that are cut are stored in the next layer grid file. There may be several layers of grid file that store unpartitioned objects, and each layer has different partitioning hyperplanes. At the maximal layer, the objects are then clipped if they intersect the partitioning hyperplane. Figure 2.13 illustrate the 3-layer grid files: objects with solid lines and points are stored in the first layer, objects with dashed lines in the second and objects with dotted lines in the third. Using multiple layers, the number of objects being clipped is reduced greatly as compared to use the clipping technique in a single layer grid file. The intersection search for the multi-layer grid file has been shown to be more efficient than that for the grid file using object clipping [SiW88].

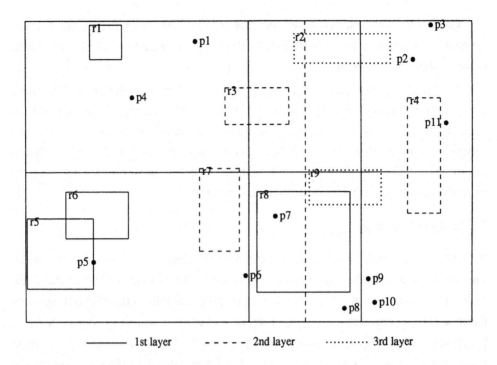

————— 1st layer   - - - - - 2nd layer   ·········· 3rd layer

Figure 2.13 Multi-layer grid files

2.5.5.2 The EXCELL Method

The EXCELL method of Tamminen [Tam82a, Tam82b] uses the same approach as that of the grid file. The major difference between the two is the way the search space is partitioned. Instead of allowing a space to be partitioned freely, the EXCELL method requires that all partitioning lines be equi-distant. When an interval is partitioned, other intervals must be partitioned as well. All grid cells cover the same amount of space, and the size of the grid directory is doubled with every split. The EXCELL method simplifies the grid partitioning operations at the expense of requiring larger directories.

In [Tam83], Tamminen suggested a hierarchical EXCELL method to alleviate the problem of large grid directories. The approach is similar to the multi-level grid file implementation proposed in [WhK85], in which each cell may be correspond to a data page or a sub-directory. However in Tamminen's hierarchical EXCELL method, overflow pages are supported to limit the maximum depth of the hierarchy. In [TaS82], the EXCELL was suggested for indexing non-zero sized objects by duplicating object identifiers.

Other file organizations based on hashing are the interpolation-based index maintenance method proposed in [Bur83, DaH85]. These methods are are k-dimensional generalization of linear hashing.

The file organizations based on hashing are generally designed for multi-dimensional point data. To use them for spatial indexing, the mapping of objects from k-dimensional space to $2k$-dimensional space or duplication of objects identifiers are generally required. However, as has been claimed in [WhK, Gut84, WhK85], although the mapping technique allows the cell methods to index multi-dimensional non-zero sized objects, it is not efficient for spatial query retrievals.

2.5.5.3 PLOP-Hashing

In [KrS88], a grid file extension was proposed for the storage of non-zero sized objects. The method is a multi-dimensional dynamic hashing scheme based on Piecewise Linear Order Preserving (PLOP) hashing. Like the grid file [NHS84], the data space is partitioned by an orthogonal grid. However, instead of using scales to define partitioning hyperplanes along each dimension, k binary trees are used for a k-dimensional space. Each internal node of a binary tree stores a $k-1$-dimensional partitioning hyperplane. Each leaf node of a binary tree is associated with a k-dimensional subspace (a slice), where the interval along the axis it associated is a sub-interval and the other $k-1$ intervals assume the intervals of the global space. Each slice is addressed by an index i stored in its leaf node. To each cell, a page is allocated to store all points that fall in the unpartitioned subspace. From the indexes stored in k binary trees, the address of a page can be computed. Following the overlapping scheme used in [Ooi87, OMS87], two extra values are stored in a leaf node to bound the objects whose centroids are in the corresponding slice along the axis the binary tree associated. Figure 2.14 gives an example for such structure.

The implementation of PLOP-hashing presented in [SeK88] assumes that the binary trees are stored in main memory. In [SeK88], it was shown analytically that it is more efficient than the R-tree [Gut84] and R$^+$-tree [SRF87] for certain distributions of spatial objects.

2.5.6 Quad-Tree Based Structures

The quad-CIF-tree [FKS81, Sam84] (where CIF denotes Caltech Intermediate Form) was proposed for representing a set of small rectangles for VLSI application. It is organized in a similar way to the region quad-tree. A region is recursively partitioned until the resulting quadrants do not contain any rectangle. During the subdivision, all rectangles that intersect with either of the two partitioning lines are associated with the subdivision

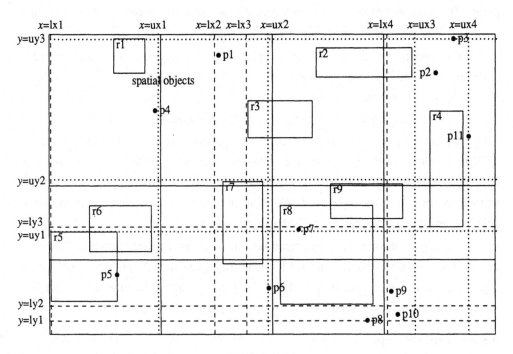

(a) PLOP-Hashing

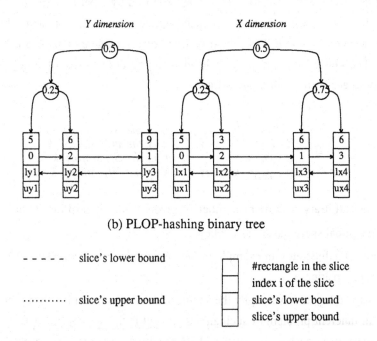

(b) PLOP-hashing binary tree

- - - - - slice's lower bound

. . . . slice's upper bound

#rectangle in the slice
index i of the slice
slice's lower bound
slice's upper bound

Figure 2.14 PLOP-hashing with binary trees for spatial indexing

point. The rectangles that are associated with a quadrant must not belong to any ancestor quadrant. It is assumed that no two rectangles overlap.

An intersection search performed on the quad-CIF-tree would begin with the root node, and examine the rectangles associated with it. The child nodes are only visited if the quadrants intersect the query region.

To insert a rectangle, each subdivision node (quadrant) is checked. If one of the axes intersect with the rectangle, the rectangle is inserted at that node. Otherwise the child node whose quadrant contains the rectangle is searched. If the quadrant does not have any child and the rectangle has not been inserted, then the process of recursive subdivision of quadrants is required. Deletion of a node is a counterpart of the node splitting during insertion, where tree collapsing is involved.

In contrast to the quad-CIF-tree representation, a point representing a rectangle may be used to store rectangles in a PR quad-tree [Sha86]. A problem with such representation is that practically an entire tree has to be searched for intersection queries, which is due to the fact that rectangles may span any parts of the space.

2.5.7 Locational Keys

A similar approach to the quad-CIF-tree was proposed by Abel and Smith [AbS83]. A quad-tree divides a space into four equal sized square subspaces. For each subspace of a quad-tree, a unique numeric key which consists of a sequence of base 5 digits can be attached. The key k for a subspace of level h can be derived from the key (k') of the ancestor subspace by the following formula:

$$
k \equiv
\begin{cases}
k' + 5^{m-h} & \text{if} & k \text{ is the SW son of } k' \\
k' + 2 * 5^{m-h} & \text{if} & k \text{ is the NW son of } k' \\
k' + 3 * 5^{m-h} & \text{if} & k \text{ is the SE son of } k' \\
k' + 4 * 5^{m-h} & \text{if} & k \text{ is the NE son of } k'
\end{cases}
$$

Here m is an arbitrary maximum number of levels in decomposition, which is greater than h. The global space has 5^m as the key.

Figure 2.15 illustrates an example of key assignment (base 5), where the maximum level of decomposition is 3.

To assign a key to a rectangle, the smallest block which totally covers the rectangle is used. An inherent problem of such an assignment is that an object MBR may be very much smaller (as a consequence of the MBR spanning one or more subspace divisions) than the associated quadrant. To alleviate the problem, a decomposition technique [AbS84] is used, where a rectangle may be represented by up to four adjacent quadrants. Rectangles B and C in Figure 2.16 illustrate the cases where two and four quadrants are

1222	1224	1242	1244	1422	1424	1442	1444
—1220—		—1240—		—1420—		—1440—	
1221	1223	1241	1243	1421	1423	1441	1443
—1200—				—1400—			
1212	1214	1232	1234	1412	1414	1432	1434
—1210—		—1230—		—1410—		—1430—	
1211	1213	1231	1233	1411	1413	1431	1433
—1000—							
1122	1124	1142	1144	1322	1324	1342	1344
—1120—		—1140—		—1320—		—1340—	
1121	1123	1141	1143	1321	1323	1341	1343
—1100—				—1300—			
1112	1114	1132	1134	1312	1314	1332	1334
—1110—		—1130—		—1310—		—1330—	
1111	1113	1131	1133	1311	1313	1331	1333

Figure 2.15 Assignment of locational keys

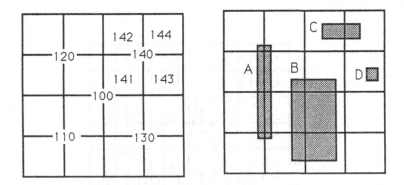

Figure 2.16 Assignment of covering nodes

used: keys 110, 130, 123 and 141 for rectangle B, and keys 142 and 144 for rectangle C. By associating each rectangle with a collection of quadrants, a better approximation of a rectangle is achieved. This form of representation duplicates the object identifiers. However, even if this approach is adopted, the size of the quadrants can be large relative to the size of the object's MBR, consider rectangles A and B in Figure 2.16.

A B⁺-tree is used to index the objects based on their associated locational keys. For an intersection search, all quadrants that intersect the query region have to be scanned. The major advantage of the use of the locational key is that B⁺-tree structures are widely supported by conventional DBMSs.

The use of locational keys has also been proposed by [Ore86]. The ordering of grid cells uses an assignment of locational keys known as Z-ordering. A space consists of 2^m × 2^m grid cells, and a rectangle is described by $[x_1, x_2]$ and $[y_1, y_2]$. A grid cell is referred to as an *element* and each of these elements has an unique z-value. The z-value of an element is obtained by interleaving the binary bits using all dimensions. For example, an element with [011, 100] ([3, 4]) is addressed by 011010 (26). In the above example, the first bit of the z-value is obtained using the X co-ordinate value. The z values of the elements trace out the path shown in Figure 2.17.

For a region containing more than one grid cells, the common first n_i bits, along each dimension i are determined. In a 2-dimensional space, all points lying inside the region have coordinates with the same n_x (along X axis) and n_y (along y axis) bit prefixes. Then the z-value of the region is constructed by interleaving the n_x and n_y bits. For example, a 2-dimensional region defined by 0010 and 0011 along X axis, and 1100 and 1111 along Y axis has 00 and 11 as common prefixes for the X and the Y axes respectively. The z-value for the region is therefore 0101 and this bit string uniquely identifies the region.

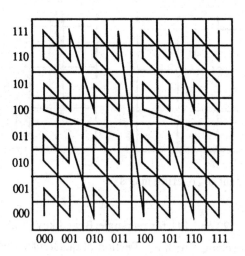

Figure 2.17 Spatial interpretation of z order

The assignment of locational keys to spatial objects is a mapping of objects from a *k*-dimensional space to a 1-dimensional space. The implementation is straight forward and conventional indexes can be used to index the objects. However, the intersection search may not be efficient.

2.5.8 Matsuyama's kd-Tree

While most kd-trees are known to be unsuitable for indexing non-zero sized objects, the kd-tree (Figures 2.18a and 2.18b) proposed by Matsuyama et al. [MHN84] is designed for 2-dimensional non-zero sized spatial objects and is based on duplication of object identifiers. To distinguish this kd-tree organization from other kd-trees, we refer to Matsuyama's kd-tree as an mkd-tree. Non-leaf nodes contain a discriminator (an ordinate), a discriminator value (an ordinate value) and two pointers each pointing to a subtree. Leaf nodes contain the address of a secondary page and a page offset. Objects belonging to a leaf node are stored in the secondary page at the given page offset. Note that multiple leaf nodes may share a secondary page. Each page contains the identifiers of objects which are partially or totally included in the corresponding subspaces and consequently, object identifiers may be duplicated in more than one page. As an example, the object *r8* in Figure 2.18 is duplicated in three different subspaces.

Each node in a mkd-tree represents a rectangular region. The root node denotes the whole map space, and a leaf node represents an unpartitioned rectangular subspace. A recursive algorithm is used to search for a data object; at each non-leaf node, the co-ordinates of the descendant subspaces are obtained and a test is performed to determine whether these subspaces intersect the query region. When a leaf node is reached, the page associated with the leaf node is fetched and spatial testing may be performed on each object.

To add an object, the object identifier needs to be inserted into all the pages that contain subspaces that intersect with the data object. It is quite common that object identifiers may be duplicated in more than one page, particularly so when the sizes of objects are large. Whenever a page overflows, the page is split with a partition being introduced along the longer side of the rectangle. The subspace is partitioned into two subspaces and the two new pages contain all objects that intersect with their subspace.

To delete an object, it is necessary to search for all unpartitioned subspaces that intersect with the data object and delete all identifiers referring to the data object. If the deletion of an object causes a page to be empty, the corresponding leaf node is marked *NIL*. The underflowed data pages are not merged.

This is one of the indexing structures adopting the replication approach. The replication approach is not suitable to store object directly in the leaf nodes as the

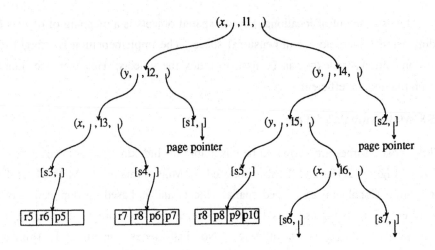

Figure 2.18a The 2-d directory for matsuyama kd-tree

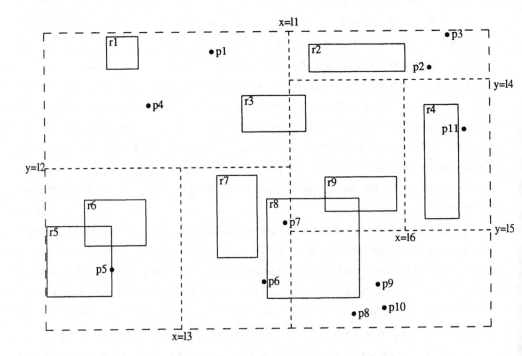

Figure 2.18b The 2-d space coordinate representation

overhead of redundant storage can be very high. Hence, object identities are usually stored in multiple leaf nodes and the objects are stored only once in an external file.

2.5.9 The 4-D-Tree

The kd-tree [Ben75] can be used to index 2-dimensional rectangular objects by mapping the objects into points in a 4-dimensional space [Ros85]. Each 2-dimensional rectangle described by $(x1, y1)$ and $(x2, y2)$, is treated as a four attribute tuple $(x1, y1, x2, y2)$. The discriminators are used cyclically and the nodes at the same level use the same discriminator. In [Ros85], the issues involved in mapping the data structure onto pages in secondary memory was not addressed. The same approach for the K-D-B-tree [Rob81] was suggested by Banerjee and Kim [BaK86]. The structure is known as the 4-d-tree.

The major problem associated with the 4-d-tree is the intersection search, which can be very costly. Suppose the search rectangle is $(qx1, qy1, qx2, qy2)$. If the discriminator of an internal node is $x1$, then the left subtree has to be searched if $qx2$ is less than $x1$, else both subtrees have to be searched. Similarly for a node whose discriminator is $x2$, where the right subtree is searched if $qx1$ is greater than $x2$, otherwise both subtrees have to be searched. The same argument applies for the Y axis.

2.5.10 The R-Tree

The R-tree [Gut84] is a multi-dimensional generalization of the B-tree, and hence the tree is height-balanced. Like the B-tree, node splitting and merging are required for inserting and deleting objects.

A leaf node contains entries of the form

$$[I, object\text{-}identifier]$$

where *object-identifier* is a pointer to a data object and I is a k-dimensional bounding rectangle which bounds its data objects.

Non-leaf nodes contain entries of the form:

$$(I, child\text{-}pointer)$$

where *child-pointer* is an address of a lower level node in the R-tree and I is a bounding rectangle covering all the rectangles in the lower nodes in the subtree. Figures 2.19a and 2.19b illustrate the structure of a R-tree and its planar representation.

In order to locate all objects which intersect a query rectangle, the search algorithm descends the tree from the root. For all rectangles in a non-leaf node that intersect with the search object, the corresponding child-pointer becomes the root of a subtree that will be searched subsequently.

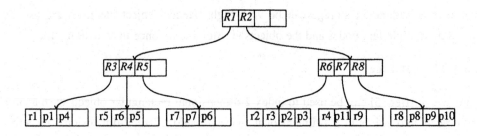

Figure 2.19a The structure of an R-tree

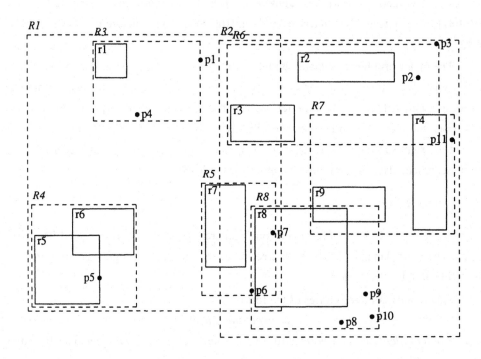

Figure 2.19b The planar representation of an R-tree

To insert an object, the tree is traversed and all the rectangles in the current non-leaf node are examined. The constraint of least coverage is employed to insert an object: the rectangle that needs least enlargement to enclose the new object is selected, the one with smallest area is chosen if more than one rectangle meets the first criterion. The nodes in the subtree indexed by the selected entry are examined recursively. Once a leaf node is obtained, a straight forward insertion is made if the leaf node is not full. However, the leaf node needs splitting if it overflows after the insertion is made. For each node that is traversed, the covering rectangle in the parent is readjusted to tightly bound the entries in the node. For a newly split node, an entry with a covering rectangle that is large enough to cover all the entries in the new node is inserted in the parent node if there is room in the parent node. Hence, splitting may propagate to the root.

To remove an object, the tree is traversed and each entry of a non-leaf node is checked to determine if the object overlaps its covering rectangle. For each such entry, the entries in the child node are examined recursively. The deletion of an object may cause the leaf node to underflow. In this case, the node needs to be deleted and all the remaining entries of that node are reinserted from the root. Similar to the node splitting, the deletion of an entry may cause further deletion of nodes in the upper levels. Thus, entries belonging to a deleted *ith* level node must be reinserted into the nodes in the *ith* level of the tree. Deletion of an object may change the bounding rectangle of entries in the ancestor nodes. Hence readjustment of these entries is required.

In searching, the decision whether to visit a subtree depends on whether the covering rectangle overlaps the query region. It is quite common that several covering rectangles stored in the same R-tree node overlap the query region, which results in the traversal of several subtrees. Therefore, the minimization of overlaps of covering rectangles as well as the coverage of these rectangles is of primary importance in constructing the R-tree [Gut84, RoL85].

A spatial index is often constructed from an existing database. The data can be pre-organized such that the index is created and optimized over a whole set of objects at one time in order to obtain better storage utilization and smaller query retrieval cost. Building a tree using preprocessing is known as static construction.

Packed R-trees [RoL84, RoL85] have been introduced to minimize the coverage and overlap of rectangles by building an R-tree statically. It has been shown that for point data, it is possible to partition points into groups such that the bounding rectangles of these groups do not overlap. However, achieving zero overlap may require rotating the orientation of the entire database, which may not be possible or beneficial. Further, zero overlap is achieveable only at the leaf level of the R-tree with static construction. For bounding rectangles associated with the non-leaf nodes, overlap is sometimes unavoidable. The main objective of the algorithm is the reduction of the storage, the coverage and overlap of rectangles, in order to improve the search efficiency.

The algorithm first orders the objects along one of the co-ordinate axes (e.g. X) using the smaller ordinate value for each MBR. The algorithm then chooses the first object from the list and using that object, selects the nearest M-1 objects to form a node, where M is the maximum number of objects or entries to be stored in a page. The process is repeated till all objects are assigned to nodes. The MBR enclosing all objects in a leaf node becomes an object at the next higher level. These objects are once again ordered and assigned to nodes in the same way. The algorithm stops when the set of remaining objects has no more than M elements; these are assigned to the root node of the R-tree.

Although the algorithm [RoL84, RoL85] uses a *nearest* function to identify the next object to be included in the current node, this function is not defined. As far as the measure of distance is concerned, there are several ways to determine the distance between two objects, for example:

(1) the minimum distance between any two points on the boundary of each object;

(2) the distance between the centroids of the objects;

(3) either (1) or (2) with respect to the MBRs of the objects;

To minimize the coverage of rectangles in a packed R-tree, the use of distance between the centroids of the MBRs is a better measure than the two nearest points because:

(1) objects that are close together based on the latter measure of distance can still have large coverage, and

(2) the coverage of both A and B are determined the maximum and minimum values of the two objects, which are the boundaries of MBRs.

A slightly different way of packing is to incorporate the enlargement criterion of the R-tree to select the next entry. That is, using the MBR of all the entries selected so far, the next entry is the one that requires the least enlargement of the MBR to cover the new entry. In Figure 2.20, rectangle b requires smallest MBR expansion, but rectangle a is nearest to the node MBR. Hence, rectangle b will be selected with the above method, but rectangle a will be selected by the packed R-tree strategy.

Figure 2.20 Selection of next rectangle

2.5.11 The R$^+$-Tree

The R$^+$-tree [SRF87] is a compromise between the R-tree and the K-D-B-tree [Rob81] and was proposed to overcome the problem of the overlapping covering rectangles of internal nodes of the R-tree.

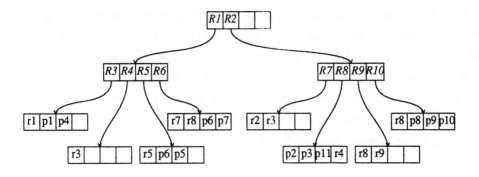

Figure 2.21a The structure of an R⁺-tree

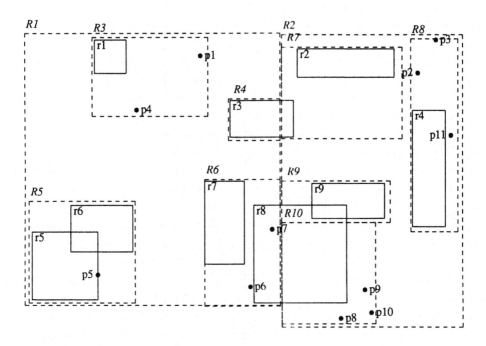

Figure 2.21b The planar representation of an R⁺-tree

The R⁺-tree structure is exactly the same as that of the R-tree, however the constraints are slightly different.

(1) Nodes of an R⁺-tree are not guaranteed to be at least half filled.

(2) The entries of any intermediate (internal) node do not overlap.

(3) An object identifier may be stored in more than one leaf node.

The duplication of object identifiers leads to the non-overlapping of entries. In a search, the subtrees are searched only if the corresponding covering rectangles intersect the query region. The disjoint covering rectangles avoid the multiple search paths of the R-tree for point queries. For the space in Figure 2.21, only one path is traversed to search for all objects that contain point $p7$. Whereas for the R-tree, two search paths exist. However, for certain query rectangles, say the left half of object $r8$, the R^+-tree is more expensive than searching the R-tree.

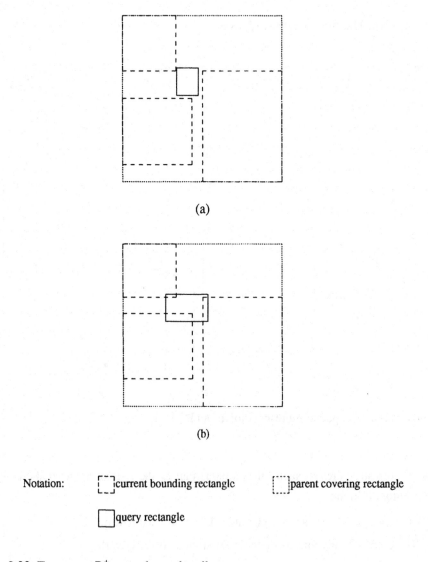

(a)

(b)

Notation: ⌐⌐⌐ current bounding rectangle   ⌐⋯⌐ parent covering rectangle

              □ query rectangle

Figure 2.22 Two cases R^+-trees do not handle

To insert an object, multiple paths may be traversed. At a node, for all entries whose covering rectangles intersect with the object MBR, the subtrees of the entries must be traversed. On reaching the leaf nodes, the object identifier will be stored in the leaf nodes; multiple leaf nodes may store the same object identifier.

Incidentally, the insertion algorithm given in [SRF87] is not complete [Ooi88, Gun88]: it does not handle the case of inserting an object into a node where the covering rectangles of all entries do not intersect with the object MBR (see Figure 2.22a). Another case not covered is one in which the MBR of the new object only partially intersects with the bounding rectangles of entries (as shown in Figure 2.22b); this requires updates of bounding rectangles. Both cases need to be handled by a smart algorithm in such a way that the two parameters, namely coverage and duplication, are minimized.

A more serious problem is that sometimes covering rectangles of entries prevent each other from expanding to include the new object. In other words, some space ("dead space") within the current node cannot be covered by any of the covering rectangles of the entries in the node, and if the new object occupies such a region it cannot be fully covered by the entries. As an example the solid rectangle in Figure 2.23 cannot be covered by the bounding rectangles (dashed rectangles). To avoid this situation, a look ahead strategy for finding the entries to include an object is necessary. The strategy must ensure that no dead space results. Alternatively, as suggested in [Sel87], the criterion for which covering rectangles are chosen to cover a new object follows that proposed by Guttman [Gut84]. When the object cannot be fully covered, one or more of the covering rectangles are split. This means the split may cause the children of the entries to be split as well, which may further degrade the storage efficiency.

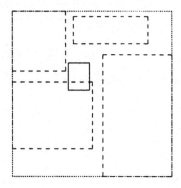

Figure 2.23 The deadlock: The input rectangle cannot be covered

During an insertion, if a leaf node is full and a split is necessary, the split attempts to reduce the identifier duplications. While the split of a leaf node may propagate upwards to the root of the tree, the split of a non-leaf node may propagate downwards to the leaves. For the example shown in Figure 2.24, two child nodes have to be split as well. Although four criteria are proposed in the splitting algorithm to improve the efficiency of the structure, no attempt is made to choose among the four. This is needed since all four criteria **cannot** possibly be satisfied at the same time. One of the criteria is to minimize the number of covering rectangles of the next lower level that must be split as a consequence, this will not guarantee that the number of rectangles that must be split overall should be minimized.

Although the problem of overlapping rectangles of the R-tree is overcome, the R^+-tree inherits many of the problems of the K-D-B-tree and the mkd-tree:

(1) Like K-D-B-tree, partitioning a covering rectangle may cause the covering rectangles in the descendant subtree to be partitioned as well. Frequent downward splits tend to partition the already under populated nodes, and hence the nodes in an R^+-tree may contain less than $M/2$ entries.

(2) Objects identifiers are duplicated in the leaf nodes, the extent of duplication is dependent on the spatial distribution and the size of the objects. To delete an object,

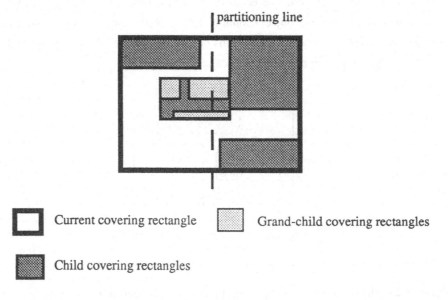

Figure 2.24 Downward split propagation

it is necessary to delete all identifiers that refer to that object. Deletion may necessitate major reorganization of the tree.

2.5.12 The Cell Tree

Based on the binary space partitioning trees [FKN80], Gunther [Gun88] proposed a *cell tree* to alleviate the overlapping bounding rectangle problems of R-trees and the "dead space" problems of R$^+$-trees. Binary space partitioning trees are binary trees that represent a recursive subdivision of a space into subspaces. The cell tree is height-balanced tree. The partitioning hyperplanes in a cell tree may not be parallel to any axis, and as a result, the unpartitioned subspaces are polyhedral. In a cell tree internal node, the bounding polygon of an entry is a convex polyhedron. The sons of each node form a binary space partition of the node. Partitions do not overlap. The fact that polyhedras may require different number of points to represent is hard for the cell tree to set a lower bound on the number of entries. An insertion may cause a node to overflow. When a page split is not possible, a node may occupy more than one page. It is nevertheless the intention of the algorithm to limit the page requirement of each node to one page.

Spatial objects are no longer bounded by MBR, but by a convex polyhedron. A convex polyheral cover of an object is composed of a set of polyhedra to better approximate irrugularly shaped spatial objects. Like the R$^+$-tree, an spatial object being represented may be stored in more than one leaf node.

2.5.13 Summary

Spatial (geometric) searching is similar to aspatial multi-key searching in as much as the coordinates may be mapped onto key attributes and the key values for each object represent a point in a k-dimensional space. However, spatial objects often cover irregular areas in multi-dimensional spaces and thus cannot be solely represented by point locations. Although a technique such as mapping regular regions to points in higher dimensional space enables point indexing structures to index regions, such representations do not help support spatial operators such as intersection and containment. Hence conventional multi-dimensional index structures may be suitable for point objects but not for non-zero sized spatial objects.

We have compiled an extensive review of structures that are suitable for indexing non-zero sized objects in a GIS. The strengths and weaknesses of each of these structures have been identified. The contributions of the study are two-fold; firstly, the study suggests many useful and relevant ideas that may serve as guidelines in designing new data structures, and secondly it allows us to compare qualitatively new structures with the existing data structures.

Three major techniques being used to augment the existing multi-dimensional point structures for non-zero sized spatial objects are described below together with their advantages and disadvantages.

(1) *Object Mapping (Transformation)*: mapping of objects from a k-dimensional space into points in a $2k$-dimensional space.

To use a point indexing structure by mapping objects from k-dimensional space into points in $2k$-dimensional space requires no alteration of the structure. However, spatial searching strategies need to be introduced. The problem with the mapping scheme is that k-dimensional objects that are spatially close in a k-dimensional space may be far apart when they are represented as point in a $2k$-dimensional space. As a consequence, intersection search can be slow. Also the complexity of insertion operation typically increases with dimensionality.

(2) *Object Duplication*: storing of object identifier in all subspaces that the object intersects.

Object Clipping: partitioning an object into smaller objects and storing the smaller objects in subspaces that totally include them.

The most important property of object duplication or clipping is that the data structures used are straight forward extensions of the underlying point indexing structures. Also, both points and multi-dimensional non-zero sized objects can be stored together in one file without having to modify the structure. However, a drawback is obviously the duplication of objects which requires extra storage and hence more expensive insertion and deletion procedures. Structures adopting this extending approach are not suitable for storing objects directly in the leaf nodes. Object identities are usually maintained in the structures and objects are stored only once in an external file. As a result, additional disk accesses are incurred. Another limitation is that the density (number of objects that contain a point) in a map space must be less than the page capacity (the maximum number of objects that can be stored on a page).

(3) *Object Bounding*: maintaining enclosing rectangles such that objects are totally included in a subspace.

Maintaining effective enclosing rectangles can be expensive. Ineffective enclosing rectangles tend to overlap and hence multiple search paths result. The approach also requires extra storage for the directory structures and hence the search may be expensive.

A classification of each hierarchical indexing structure introduced in this chapter is given in Table 2.2.

In summary, the kd-tree is extended to index spatial objects, by either mapping non-zero sized objects into points in higher dimensional space [BaK86], or by object duplication [MHN84]. Both extensions are simple, but not efficient in terms of searching and/or storage. A number of grid file implementations have been proposed [Hin85, NHS84, Tam83, WhK85]. As stated in [WhK85], the two disk access principle of grid files is only possible if the directory is stored as an array. The implication of using an array is that the size of the array is doubled for each directory split. An additional problem of the grid file stems from the mapping of non-zero sized objects. R-trees [Gut84] are generalization of the B-tree data structures. Even though the R-tree has the problem of overlapping rectangles, it is generally a well defined and reasonably efficient structure. The R^+-tree [SRF87] was proposed to solve the overlapping rectangle problem of the R-tree, and for certain classes of spatial objects, it has been shown [FSR87] to be more efficient than the R-tree. For non-point data, the problems of identifier duplication, storage underflow and extra storage overhead arise with the R^+-tree. Based on the grid-files and the object bounding method used in [Ooi87, OMS87], the PLOP-hashing was proposed to index non-zero sized objects. Although it has been shown anlytically to be

Table 2.2 Indexing techniques used by structures

Structures	Spatial Techniques
Grid Files	Object Mapping
EXCELL	Object Mapping or Duplication
PLOP-Hashing	Object Bounding
locational keys	Object Mapping
R-Trees	Object Bounding
R^+-Trees	Object Bounding and Duplication
Cell-Trees	Object Bounding and Duplication
4-D-B-Trees	Object Mapping
mkd-Trees	Object Duplication
skd-Trees	Object Bounding

more efficient than both the R-tree and R$^+$-tree for simple data, it remains to be shown for general cases. Further, the access of the binary-tree directory has never been taken into account.

Features of one structure may be used to complement the weaknesses of the others. For example, the recursive decomposition of space techniques of grid files or quad-trees may be used in R-trees to determine the expansion of enclosing rectangles and hence the insertion of objects. Objects that are much larger than average objects or that cause severe overlap may be stored in the internal nodes or seperately. However, new problems may be introduced.

2.6 Optimizations

2.6.1 Introduction to Query Optimization

One of the major considerations in designing and implementing a database system is efficient retrieval of stored information. In systems with low level query languages in which physical data access paths must be specified, the efficiency depends on how well users can express their queries. On the other hand, high level query languages (e.g. as supported by relational database systems) allow users to express queries without reference to access paths, and so the system, not the users, is responsible for the efficient processing of queries. For these reasons, a great deal of effort has been focused on the automatic optimization of queries for relational databases. Even if the strategy used may, sometimes, not be the best strategy, it is customary to refer to the process of *finding an efficient strategy for executing a query* as **query optimization**.

What to optimize, one may ask? An immediate answer is the response time, which is the amount of time required for a query processor to answer a query. Another optimization objective is the total cost, which is dominated by the secondary storage access cost and the somewhat less expensive computation cost. Running the optimizer also takes time, and in pathological circumstances the cost that is saved may be less than the cost of finding the optimal plan. Consequently, some well known systems (e.g. System R [ABC76]) opted for a simple and efficient scheme for computing a close to optimal execution strategy.

One strategy adopted in finding an evaluation plan for a query is based on certain built-in rules for ordering the execution of various operations. A more general approach is to generate a set of execution plans, which is small enough to be feasible and yet large enough to contain a near optimal plan, and to choose one among all generated plans using some cost formulae. The latter approach is deemed preferable as preprocessing strategies such as logical transformations can be added to.

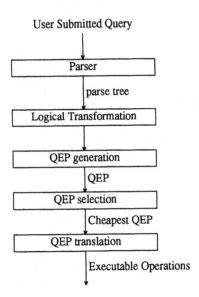

Figure 2.25 A general optimization strategy

A general top-down optimization strategy consists of the following steps: parsing, logical query transformation, evaluation plan generation, plan selection and plan translation (Figure 2.25). The *parser* checks the syntactic and semantic correctness of the query, and casts the query into an internal representation. The optimizer then standardizes the query by transforming it into a canonical representation which may be simplified to remove redundant or needlessly replicated components. Thereafter, the optimizer uses the *meta-database* (e.g. data dictionary) to generate all possible efficient *Query Evaluation Plans* (QEPs). A QEP consists of a sequence of elementary operations, e.g. join, restrict and project, etc. The plan is then augmented with the details of available lower level operations (e.g. relation scan, indexed scan, join method) and organization of data at the file level. Using cost estimation methods, the best or the cheapest QEP is determined. The final task is to translate the QEP into executable instructions.

As mentioned in Chapter 1, access methods play an important role in efficient query processing, where the final query execution plan depends critically upon the existence of appropriate paths. The study of indexing structures is required as a part of optimization to select an appropriate and efficient structure to facilitate the query retrieval. The cost estimation is essential to determine whether the use of supported indexes can speed up the query retrieval. In some instances, the use of index structures may not be beneficial at all. The point we are making here is that the study of indexing structures is an

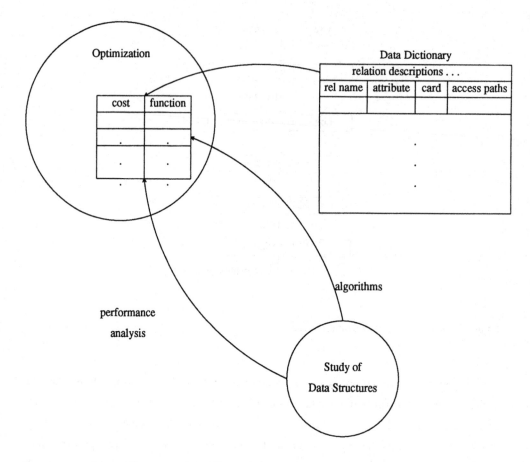

Figure 2.26 Conceptual view of optimization problem

essential prerequisite for solving query optimization problems, although the study is always done independently. Figure 2.26 illustrates the inter-connection of the two issues.

In the remainder of this section, we study optimization techniques proposed in the literature, which provide the necessary background to the hybrid query optimization techniques required for a GIS supporting GEOQL. Besides, many designers still learn the basic optimization from that used in System R or earlier systems, hence it is essential to look at more sophisticated and advanced optimization strategies.

2.6.2 Definitions

We start with some terms, notational conventions and the assumptions used throughout this thesis. Additional terminology is defined in the later parts of this thesis as required.

The reader is assumed to be familiar with relational databases [Dat86, Gra87]. The term R_i denotes a relation, which can be either a base relation or a temporary relation, and S_j is a set of relations. A predicate $P_k(S_j)$ is a condition defined over a set of relations S_j. When the identity of the relations is not important, only P_k is used. We use $Q_m(S_j)$ to denote a query which is defined over the relations in S_j.

We say two queries, $Q_a(S_a)$ and $Q_b(S_a)$, are **equivalent** if and only if their execution produce the same results. A subset A of B is said is be a set of **canonical forms** for B if and only if every object β in B is equivalent to one object α in A. The object α is said to be the canonical form for the object β.

Two modules that are often associated with a high level description of a database system's architecture are the *frontend* and the *backend*. Of course, within each module, there are many complex components such as the optimizer, concurrency and recovery controllers, etc. A frontend usually refers to a user interface unit, the main task of which is to assist *end-users* using the database and developing database applications. Most frontends in commercial systems are either *forms-based* and screen oriented or support a non-procedural query language (e.g. SQL or QBE). The backend provides most of the data management functions such as data definition, data security, data manipulation, access methods, semantic integrity subsystems, etc. In this thesis, we consider the backend to be a query processing unit, which finds an efficient query execution strategy for a given query and executes it.

2.6.3 Optimization Strategies

Disregarding interface languages, query optimization and execution requires a query to be represented in some internal form. It is highly desirable that the structure (usually a *query parse tree* or *graph*), in which the semantics of the query are captured, should not deter any subsequent optimization choices and should allow an easy transformation. The representation of the trees may be using multiway trees, binary tree and graphs [Knu73]. Whatever representation is used, the parse tree must be simple yet powerful enough to capture the full semantics of the query.

As an example, Figure 2.27 ($X(\rightarrow R)$ is used to indicate that X is an instance of R) illustrates the parse tree for the GEOQL query given in Section 2.2.4, where a multiway tree is used. A special part of the query tree, the subtree in which the qualification is described, is called the and/or predicate tree, where the logical transformation is usually performed.

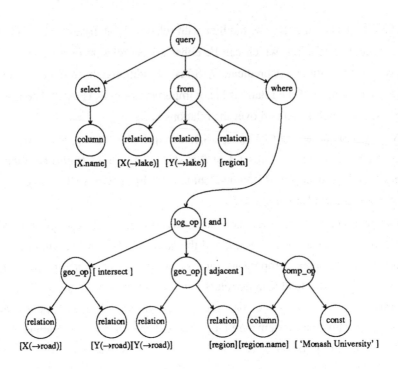

Figure 2.27 A GEOQL parse tree

2.6.3.1 Query Transformation

Much effort has been focused on logical query transformations which are used as a means to improve query execution efficiency by preprocessing queries before generating the QEPs. The proposed logical transformation techniques may fall into one or more of the following classifications:

(1) standardization,

(2) simplification, and

(3) amelioration.

An extensive survey of earlier optimization schemes can be found in [JaK84].

2.6.3.1.1 Standardization

To be able to detect and remove any redundancies of a query, we must have a standard form of representation of the query. Table 2.3 specifies transformation rules to standardize the representation of query. Two well known standard forms of representation are *Conjunctive Normal Form* (CNF) and *Disjunctive Normal Form* (DNF). A predicate in CNF is of the form

$$(P_1 \text{ or } ... \text{ or } P_k) \text{ and } ... \text{ and } (P_m \text{ or } ... \text{ or } P_n),$$

and a predicate in DNF is as follows:

$$(P_1 \text{ and } ... \text{ and } P_k) \text{ or } ... \text{ or } (P_m \text{ and } ... \text{ and } P_n).$$

Each of these two forms has its own advantages; the use of DNF enables independent optimization of query components [BGW81], and CNF allows decomposition of a complex query into a set of simpler queries [WoY76].

2.6.3.1.2 Simplification

Queries with different syntactic forms may have the same semantics, with the expression of one being possibly more efficient than the other. One source of difference between two semantically equivalent queries is their degree of *redundancy*. One of the preprocesses involved in the optimization converts a query into an equivalent but simpler form [Hal74, JaK84], such that the simplified query should not contain superfluous predicates.

Simplification of the predicate tree can be obtained by:

(1) Recognition and removal of double negations;

Table 2.3 Transformation rules

Expression	Equivalent Form
Commutative Rules	
$and(P_1(S_1), P_2(S_2))$	$and(P_2(S_2), P_1(S_1))$
$or(P_1(S_1), P_2(S_2))$	$or(P_2(S_2), P_1(S_1))$
Associative Rules	
$and(P_1(S_1), and(P_2(S_2), P_3(S_3)))$	$and(and(P_1(S_1), P_2(S_2)), P_3(S_3))$
$or(P_1(S_1), or(P_2(S_2), P_3(S_3)))$	$or(or(P_1(S_1), P_2(S_2)), P_3(S_3))$
De Morgan's Rules	
$not(and(P_1(S_1), P_2(S_2)))$	$or(not(P_1(S_1)), not(P_2(S_2)))$
$not(or(P_1(S_1), P_2(S_2)))$	$and(not(P_1(S_1)), not(P_2(S_2)))$
Distributive Rules	
$or(and(P_1(S_1), P_2(S_2)), and(P_1(S_1), P_3(S_3)))$	$and(P_1(S_1), or(P_2(S_2), P_3(S_3)))$
$and(or(P_1(S_1), P_2(S_2)), or(P_1(S_1), P_3(S_3)))$	$or(P_1(S_1), and(P_2(S_2), P_3(S_3)))$

(2) Reduction of the predicate tree based on idempotency;

(3) Recognition of equivalent subexpressions and removal of the associated redundancy.

The transformation rules are defined in Table 2.4. Earlier work in this area was started by Hall [Hal74, Hal76], who proposed a bottom up simplification algorithm to simplify the predicate tree.

2.6.3.1.3 Amelioration - Algebraic Manipulation

A query with no redundant predicate may not be in its most efficient form in terms of operations involved, and hence it may not be unique semantically. Such a query can be further improved by grouping predicates (relational calculus) or operations (relational algebra) involving the same relation(s) together.

A string of projections involving the same relation can be grouped into a single projection as follows:

$$\pi_{A_n} (...(\pi_{(A_2 \cdots A_n)} (\pi_{A_1 \cdots A_n}(R)))...) \Rightarrow \pi_{A_n}(R).$$

Table 2.4 Simplification rules

Expression	Equivalent but Simpler Form
Double Negation Rule	
not(not(P(S)))	P(S)
Idempotency Rules	
or(P(S), P(S))	P(S)
and(P(S), P(S))	P(S)
or(P(S), not(P(S)))	True
and(P(S), not(P(S)))	False
or($P_1(S_1)$, and($P_1(S_1)$, $P_2(S_2)$))	$P_1(S_1)$
and($P_1(S_1)$, or($P_1(S_1)$, $P_2(S_2)$))	$P_1(S_1)$
or(P(S), True)	True
and(P(S), True)	P(S)
or(P(S), False)	P(S)
and(P(S), False)	False

The same principle can be applied to a sequence of restrictions, and the transformation rule is defined as follows:

$$\sigma_{A_1 \text{ op } x_1}(R) \text{ log\_op } ... \text{ log\_op } \sigma_{A_n \text{ op } x_n}(R) \Rightarrow$$

$$\sigma_{A_1 \text{ op } x_1 \text{ log\_op } ... \text{ log\_op} A_n \text{ op } x_n}(R).$$

where op and log_op are arbitrary comparison and comparative operators.

Two obvious advantages in combining the operations involving the same relation are: the relation is only scanned once, and the probability of being able to use an existing access path (index) over a particular column is maximized.

Although relational algebraic expressions include a procedural component, the operations may be reordered [SmC75] to produce a semantically equivalent but more efficient sequence of operations. The cost of evaluating a query depends in part upon the size of the intermediate partial results. Indeed, intermediate results may sometimes be required to be stored as temporary relations, which will be read into main memory again when required. As such, minimization of the size of the intermediate results is of primary importance. Constructive operations such as join and Cartesian (cross) product, which increase the size of partial results, should be delayed. Operations such as restriction and projection should be moved forward over joins and Cartesian products so that selective operations can be performed as early as possible. Pipelining should be applied whenever feasible so as to avoid writing and reading of temporary relations to and from secondary storage.

Often, users tend to visualise the evaluation of a query as firstly forming the Cartesian product of all required relations, and subsequently evaluating each predicate. The algebra query in Figure 2.28 is a case in point, in which the early join would potentially create a huge temporary relation; this could be avoided by performing the selection and restriction first. The transformed query graph is as shown in Figure 2.29a. Further reduction of the size of the intermediate relations is possible, if redundant attributes not required for further operations can be discarded. Figure 2.29b shows the final sequence of operations for the query in example 2.1. Notice that the relation Employee could be projected on (job#, name, position) before the join, however we have assumed the existence of an index over the attribute job# of Employee; this index may be used for efficient evaluation of the join operation nevertheless, prior projection of Employee would mean the index could no longer be used.

Further performance gains from parallel evaluation is feasible in this case by pipelining the partial result of the selection on the Job relation to the projection on the attribute job#. In such case, no extra processor is required because the selection and projection can in fact be performed at the same time. In general, if we can achieve extra CPU-I/O overlap and have spare CPU capacity, then pipelining is synchronized to avoid I/O for intermediate results.

Example 2.1 A Relational Algebra Query.

Database: Employee(emp#, name, position, job#)
        Job(job#, dept, room)

Query: Find the name and position of employees who work
        in the EDP department.

Relational Algebra Query:

$\pi_{\text{name,position}} (\sigma_{\text{dept='DEPT'}} (\text{Employee} \bowtie_{\text{job\#=job\#}} \text{Job}))$

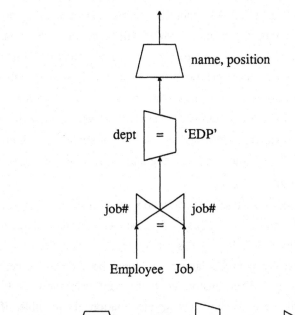

Figure 2.28 The query graph for a relational algebra query

In the context of relational calculus, an analogous optimization of the order of evaluation, namely detachment of one-variable predicates (restrictions), is introduced by [WoY76, YoW79] to restrict the relations as early as possible. The main idea is to decompose a query into smaller components such that the overlap between the components is constrained to one common tuple variable. However, Goodman and Shmueli [GoS82] showed that some expressions which contain cycles cannot be reduced

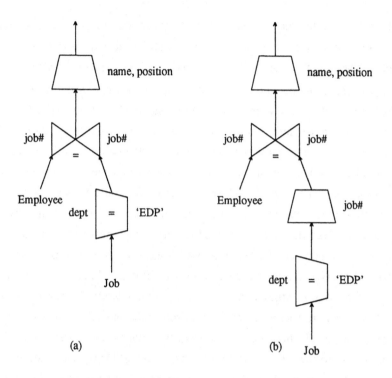

Figure 2.29 Two transformations of the query from example 2.1

in this manner. A subclass of the queries, tree queries, which contain no cycles, has been identified to be always reducible. Such cycles can be detected by a tree membership algorithm proposed by [YuO79], and the cycles can be removed by the use of functional dependency [KaY83] or tuple substitutions [WoY76].

Adding additional restriction predicates under logical implication is another way of reducing the cardinality of the operand relations. For example, the predicate ($R_1.A >$ $R_2.B$ and $R_2.B > 100$) implies $R_1.A > 100$ can be conjuncted to the predicate. The main purpose is to reduce the cardinality of qualifying tuples from R_1 as much as possible before evaluating the inequality join.

2.6.3.1.4 Semantic Optimizations

Semantic Integrity in a database context refers to the accuracy, correctness or validity of data [Dat86]. Consistency and correctness of data in a database can be achieved by enforcing semantic integrity constraints during updates to the database. The constraints are *assertions* that the committed data must satisfy at any instant. Hence, the response to a retrieved query is also consistent with the constraints imposed on the database. Making use of the semantic knowledge, semantic query optimization has emerged as a preprocessing strategy for conventional optimizations to help processing queries more efficiently. These methods are collectively known as *semantic optimization.*

Preprocessing optimization schemes using semantic constraints to improve the query processing efficiency were formally proposed in [Kin81] and [HaZ80]. The major difference between semantic optimization and conventional optimization is the use of semantic constraints and knowledge of indexes and links to introduce extra joins and restrictions, and eliminate redundant restrictions whenever the resultant query may speed up the evaluation. While in QUIST [Kin81] integrity constraints are used as semantic transformation rules, subexpressions of queries are replaced by semantically equivalent expressions in [HaZ80]. Graphs are used to derive canonical representations of queries in [ShO87]. The approach [ShO87] proposed consists of three phases: in the first phase, the canonical query graph is obtained; in the second phase, integrity constraints are used to insert new restriction edges; and in the final phase, redundant edges are removed. The method however is restricted to selection introduction and elimination [SSD88]. While the cost of query evaluation may be reduced, the total cost of query processing may be exacerbated due to the increase in search space. In [SSD88], search stopping criterion during the introduction of predicates are proposed to minimize the overall query processing cost. A more recent study of semantic query optimization is by Pang et al. [PLN89] in the context of object-oriented databases. A constraint classification approach is proposed to classify constraints based on the referenced objects. The classification scheme is used to reduce the number of constraints that need to be considered during the optimization process. Each restriction clause in the query is assigned a tag to indicate whether it is beneficial, imperative or redundant. The removal of an imperative clause of a query will change the semantics of the query, hence it must be left untouched. A beneficial clause can be removed, but its presense will improve the query efficiency. A redundant clause must be removed as it has no meaning to the query and hence the removal will only increase the efficiency. A simulation study performed shows that in generally for a large database the query cost is greatly reduced.

The use of functional dependencies and other integrity constraints (e.g. referential integrity) can be potentially useful in improving the efficiency, which is not well understood at the moment.

2.6.3.1.5 SQL Nesting Eliminations

One of the most powerful features of SQL is the ability to nest queries, which partly provides the relational completeness [Cod71a] and the power of the language. However, the processing of nested queries using nested evaluation schemes (e.g. System R [ABC76]) can be costly. Hence a method has been proposed [GaW87, Kim82] to transform a nested query into a partially ordered set of unnested queries.

The System R approach in processing nested queries, where the nested queries do not reference any relations of the outer block, is to first evaluate the inner block independently of the outer block. The result of evaluating the inner block, which contains no predicates referencing relations of the outer block, is either a constant or a list of constants. Having evaluated the inner block, the resulting query is no longer nested and hence, can be processed in a normal way. This method completely processes the inner query block for each tuple of the relation of the outer query block if the query block contains a join involving a relation of the outer block. Inevitably, this will incur expensive rereading of relations in the inner block, unless the inner relations are small enough to fit in the memory.

In [Kim82], nested queries are classified into five different types according to the type of predicates and nesting. All types can be transformed into semantically equivalent un-nested queries. With the use of temporary relations, which are created by auxiliary un-nested queries, these transformed un-nested query blocks can be processed much more efficiently. Using these algorithms [Kim82], a query of depth n can be transformed into a query of depth n - 1 plus zero or more auxiliary un-nested queries. Transformation of an arbitrary nested query can be obtained by applying the algorithm recursively.

Some minor deficiencies (with respect to the semantic correctness) of the Kim algorithm have been identified and appropriate modifications suggested [GaW87].

2.6.3.2 Query Evaluations

A technique which cannot be divorced from efficient processing of all queries is the optimization of all primitive operations, namely restrictions, projections and joins. A naive method for evaluating a restriction is to scan the relation reading and evaluating the condition for each tuple. This approach is costly when the relation is large, and often existing indexing structures and sorting may be employed to improve to reduce the number of tuples retrieved and tested.

Two commonly used joining methods are: the *nested iteration* (loop join) method and the *merge* method [BlE77, SAC79]. The former is order-independent while the latter is order-dependent. A nested iteration of two relations, R_1 and R_2, involves scanning the whole relation of R_2 (R_1) for each tuple of R_1 (R_2). The join can be augmented by the use of index(es) on the join attribute(s) of the inner relation [Klu82]. The merge method involves sorting the relations in the ascending order of the join attribute(s) if the relations are not already in sorted order. Then both relations are scanned synchronously, by which the complete join can be constructed in a single pass, and merged according to the join conditions.

Many other approaches intended to improve the join operations have been proposed; these include the use of *link* [Hae78, JaS82], prejoining of relations [Mis82], hashing [Bra84], and the use of join indexes [Val86, Val87].

2.6.3.2.1 Query Execution Plans

Having transformed a query into an equivalent and more efficient form, the query optimizer uses the existing storage structures, access paths, physical data representations (sort order etc.) and a cost model to generate all the reasonable Query Evaluation Plans (QEPs) for a query. By and large, the approach consists of the following steps:

(1) Formulation of all *plausible* QEPs.

(2) Finding of QEP from (1) with minimum *estimated* cost.

Although many schemes have been proposed (e.g. [ABC76, RoR82, SAC79, Yao79, YoW79]), they differ mainly with respect to the heuristics used to determine *what* is *plausible*. More heuristics are required for a fewer number of QEP being generated; on the other hand, a more relaxed heuristic would generate a larger set of QEP. It is conceptually true that the latter may yield a better or an optimal QEP, at the expense of a more expensive search space.

The approach advocated by Smith and Chang [SmC75] is the first strategy mentioned in Section 2.6.1; only one QEP is generated. On the contrary, Yao [Yao79] generates all possible QEPs for a two variable query, which may be very costly for complex queries. Many researchers [ABC76, RoR82, SAC79, YoW79] adopt both heuristic formulation and selection of QEP to reduce the search space.

Recently, Ioannidis and Wong [IoW87] proposed an algorithm which is based on the simulated annealing process to optimize a query involving a huge number of relations or a recursive query [BaR86] that potentially entails a large access plan space. The simulated annealing is a probabilistic hill climbing algorithm, which uses the greedy approach. Therefore, it is not suitable for conventional query optimization [IoW87] where the QEP space is not large.

When a new primitive operation is required to support the evaluation of queries, the QEP formulation module needs substantial modification. As such, a rule-based QEP generation technique, based on term rewriting [Hue80] techniques and functional programming, has been proposed in [Fre87b] to formulate different programs (QEPs) from the algebraic form of a query.

To select the best access plan, a cost model of storage structures and access operations is required. A cost model usually involves components for CPU usage [MaL86, SAC79], secondary storage accesses and main storage demand [Kim82]. One influential factor which is common to these three is the size of the operand relation and/or the size of the intermediate results. The size of base relations may be obtained from the meta-database, but the size of the intermediate results must be estimated. The extent to which this estimation is accurate depends on the assumptions such as the availability of higher order statistics (e.g. selectivities of all attributes [Yao79]), distribution of attribute values, types of queries [SAC79] etc. As stated in [JaK84], there are no generally accepted formulas for estimating the size of intermediate results.

Once the least cost QEP is selected, it is translated into a sequence of executable instructions which will be executed by the physical database processor. Such a processor in System R is known as the Relational Storage System (RSS) [ABC76], where the QEP is translated into an assembler program with embedded calls to the RSS. In [FrG86, Fre87a], a QEP is transformed into an iterative program in a high level procedural language.

2.6.3.3 Optimization in System R

System R is one of the earliest prototype relational systems developed by IBM at the San Jose Laboratory [ABC76]. Many of its query processing features may be found in the later commercial products such as SQL/DS [Dat86] and DB2 [CLS84, Dat86 Chapter 4, Haj84, IBM84]. The query language supported by System R is IBM's dialect of SQL.

A user-level SQL query, may consist of several queries, and each query may be nested. To execute a query, system R first decides the order of the queries to be executed, the order is however mandatorily determined by the nested order in the case of nested queries. Then it optimizes each query block individually.

System R maintains a rather static set of statistical information describing the status of the database. This statistical information is not updated after each database update, instead it is recomputed periodically by the execution of a special routine. The approach aims to minimize the cost of maintaining the statistical information, but as a consequence, the optimization may be based on obsolete statistical information. Furthermore, System R uses a fairly simple cost estimation scheme. A more detailed extension has been developed in its distributed version, System R^* [MaL86].

A general approach discussed in Section 2.6.3.2 is adopted. For each relation, an access method [SAC79] is chosen to minimize the cost of QEP. However, given a set of join terms, once a joining sequence is determined, the nested join is the only consideration. For example, for the following joins,

$$R_1 \text{ JOIN } R_2 \text{ JOIN } R_3 \text{ JOIN } R_4$$

if the following sequence is determined,

$$((R_2 \text{ JOIN } R_1) \text{ JOIN } R_3) \text{ JOIN } R_4.$$

even if an alternative join sequence which is semantically correct and/or more efficient such as below would not be considered.

$$(R_2 \text{ JOIN } R_1) \text{ JOIN } (R_3 \text{ JOIN } R_4)$$

Although the nested joining method of multiple relations allows pipelining, this is only plausible if there is extra CPU capacity.

2.6.3.4 Optimization in INGRES - Query Decomposition

A complex query, which involves a great number of relations and predicates, may be hard to optimize. A method to solve such a problem is to decompose a complex query into a sequence of simpler subqueries. A well known decomposition method proposed in [WoY76, YoW79] is to decompose a calculus-based query (QUEL) in a prototype database system called INGRES. The prototype system was developed at the University of California, Berkeley, from which the commercial INGRES [RoS86] and a more recent extensible system known as POSTGRES [StR86, StR87] evolved.

In addition to using the principle that restriction should be performed as early as possible. INGRES introduces the idea of decomposition of Cartesian products and joins. In essence, it has two overall objectives; one is to obtain the final answer by assembling comparatively small intermediate results, and another is to minimize the number of tuples scanned.

Two major operations involved in the decomposition process [WoY76, YoW79] are the *tuple substitution* and the *detachment*. The tuple substitution is substituting attribute values, one tuple at a time, for one variable of the n-variable query. In general, it can be described as:

$$Q(R_1, R_2, ...) \rightarrow \bigcup_{x \in R_1} Q'(x, R_2, ...)$$

The detachment is removing a component of the query that has just one variable in common with the rest of the query:

$$Q(S) \rightarrow Q_1(S_1) \, Q_2(S_2): |S_1 \cap S_2| = 1 \text{ and } S_1 \cup S_2 = S.$$

These two operations suffice to decompose any query completely. Tuple substitution for a single variable incurs the cost of processing the remaining portion of the query times the cardinality of the substituted relation. Hence the detachment is performed before the tuple substitution, and only when no further detachment is possible tuple substitutions are considered.

2.6.4 Global Optimization Strategies

Given a set of queries, a conventional query optimizer will optimize them sequentially, that is, one query at a time. Of course, the cost for processing a given set of queries in this manner is the sum of the cost in processing each individual query. One way to reduce the cost is to process several queries simultaneously, sharing the common tasks thereby avoiding redundant page access. An instance of sharing common tasks was first used by Hall [Hal74, Hal76] in finding common subexpressions within a single query. Grant and Minker [GrM81] proposed a two stage optimization procedure: the first stage is known as *Preprocessor*, which obtains information on the available access structures at the compile time; the second stage is the optimizer which groups queries and executes them as a group. Kim [Kim85] proposed a two-stage approach similar to [GrM81], and the unit of sharing is the relation being referenced. In [Jar85], Jarke described methods to detect and isolate common subexpressions in queries expressed in relational algebra, domain relational calculus and tuple relational calculus. A hierarchy of algorithms suitable for global optimization is exhibited and analysed in [Sel86]. The simulated annealing [IoW87] technique mentioned before is also applicable to global optimization. To sum up, two common techniques used in global optimization are:

(1) Decomposition of a query into a set of subexpression,

(2) Identification of common subexpressions and applicable access paths.

2.6.5 Extensible DBMS and Resident Memory Systems - Current Trends

New implementation techniques and new capabilities for database systems are being proposed at an impressive rate. Two current trends pertaining to the design and development of DBMS are extensible DBMS and memory resident systems.

Conventional DBMS are not suitable to support new applications such as CAD/CAM and geographic data analysis. This leads to the current trend of designing extensible DBMS [BaM86]. The main objective of extensible DBMS is the ability to support new data types which require new operators, indexing structures and special feature procedures without much implementation effort. Several ongoing projects which are still at an experimental stage are EXODUS [Gra87, GrD87], GENESIS [Bat86], GemStone [MSO86], MOODS [LNN90], ORION [KBC87], POSTGRES [StR86, StR87] and PROBE [BaM86, OrM88] (to name a few). Two approaches are adopted here: one is to extend the relational database model with new features (eg. POSTGRES), the other is to abandon the relational model in favor of object-oriented concepts (eg. ORION). The object-oriented database [ABD89] is the confluence of two seemingly disparate fields, namely databases and programming languages, to provide a more natural and powerful modeling capablity [Ngu90, NgO90] and software reusability [ABD89]. Although object-oriented databases are very rich in semantics, their performances have so far been not efficient [LNN90]. Majors areas that affect the performances are the storage structures, query processing strategies and concurrency control strategies. These are largely due to the support of large complex objects — objects that consist of component objects.

While all these projects may have different implementation, one of the common objectives is the capability of supporting an optimizer [POL90] or multiple optimizers for different data models [GrD87]. In EXODUS [GrD87], the optimizer can be generated by a rule-based system. The approach adopted by [Fre87b] advocated more or less the steps illustrated in Figure 2.30.

The specification of transformation rules describes the generation of QEPs from initial queries; these transformation rules make use of existing indexing structures and operations to generate different QEPs. As proposed in [Fre87b], these rules are translated into transformation procedures. The search strategy to prune the search space of QEPs and the cost criteria used in choosing the best QEP are then merged with the transformation procedures to generate the optimizer. Simple conventional optimizers generated by the EXODUS extensible system have been shown to be as effective as those of conventional database systems [GrD87]. Many questions as put forward by Freytag

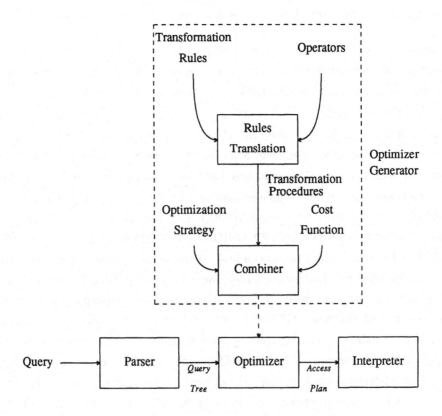

Figure 2.30 The optimizer generator of an extensible DBMS

[Fre87b] remain open: a major one, is it possible to show that a given set of rules generate only valid QEPs?

Another current trend is the main memory DBMS, which is predicted to attract more attention as the cost of chips decreases while memory capacities increases. With memory sizes growing as they are, it is quite likely that the databases will eventually entirely fit in main memory. The memory based DBMS requires different attention from the conventional DBMS receives. In the context of query optimizations, different techniques for primitive operations and cost criteria are required. The cost is no longer dominated by the page accesses, instead, the comparison cost which are usually ignored in the conventional system is now an imperative factor. The availability of large main memories makes hash-based algorithms much more efficient than the sort merge join. One such algorithm is called *Hybrid algorithm* [DKO84], which partitions a join into a set of smaller joins and then use hashing to join the tuple sets [Sha86]. The method is based on the earlier sort-merge join method proposed in [KTM83]. The simulation reported in [DKO84] indicates that the hashing is more efficient than the sort-merge join. In a conventional database, performing a projection before a join usually is beneficial, which is not the case in a memory resident database, because a list of pointers pointing to the tuples is stored as the intermediate results [LeC86, Leh86]. Further, an index for a memory resident database stores, instead of actual values, pointers through which the attribute values can be extracted.

2.6.6 Optimization in GIS

Although many external interface languages and data models have been proposed for GIS or CAD/CAM systems, little attention has been paid to the optimization of hybrid queries resulted from the extended languages. Although a lot of work has been done on spatial search structures and integrating them into query languages, to our knowledge, no one has addressed the query optimization in the context of extended database systems.

The earlier systems that store each line segments of a polygon as a tuple presumably optimize queries in the same way as a conventional DBMS. But the reconstruction of polygon from these line segments and the evaluation of spatial predicates might not be feasible. In CORGIS [Abe86], evaluations of spatial predicates are always done before the conventional SQL predicates. The resulting objects identifiers are then used as the SQL containment predicate (predicate in). The optimization problems are not addressed for PSQL [RoL84, RoL85, RFS88], and for extended QUEL [OFS84].

2.6.7 Summary

We have given a review of a variety of optimization techniques within the framework of relational calculus. Most of these techniques adopt the top-down approach to solve the optimization problems.

One of the major tactics used is logical transformation, in which queries are transformed into a more efficient form. A promising direction in this area is the use of semantic information to improve a query.

Rule-based methods are used increasingly to provide more intelligent and modular query processing strategies. To accommodate different and unconventional data models, research into extensible systems has been initiated recently. Collectively, these techniques form a solid base for our work in finding an optimization strategy for a system in which unconventional complex data structures are required.

2.7 Conclusion

A brief overview of GIS was given in this chapter. Several known interface languages, which augmented existing well known query languages, were also discussed. GEOQL [SMO87] is proposed as the external interface language for a GIS in this project.

A query cannot be processed efficiently without efficient indexing mechanisms. Like conventional databases, geographic databases require indexing structures to provide fast access to data based on proximity. Indexing structures suitable for spatial objects were reviewed, and their strengths and weaknesses identified.

An overview of optimization strategies was presented. The current trend is oriented towards a top-down approach, where solutions for a subset of problems are gradually incorporated to a general optimization strategy. Much of the work has been focused on preprocessing transformation techniques for improving the efficiency of query processing by restructuring queries to avoid expensive nesting evaluations, redundant expressions, and to take advantage of links and indexes.

The trend of using rule-based approaches in query optimization and DBMS design is a result of combining research achievements from the fields of artificial intelligence (logic programming in particular) and databases, and its impact is well recognized by several recent new conferences (e.g. Roles of Artificial Intelligence in Data Bases, Conference on Expert Database Systems). The integration is motivated by the fact that both logic programming and relational calculus share the same underlying mathematical model.

More recent trends in designing extensible DBMS and memory resident database systems have gained momentum, and will continue as memory cost decreases and database techniques are applied to new application areas.

To summarize this chapter, although conventional methods excluding the less understood extensible techniques have, repeatedly, proved to be an efficient tactic in queries processing, these methods cannot be used to efficiently evaluate spatial queries. In general the requirements in implementing a GIS in a database context are:

(1) A powerful interface language that is capable of expressing both spatial and aspatial selection criteria. This is usually an augmentation of a well known conventional language.

(2) An efficient spatial indexing structure is necessary as a speed-up device to provide fast access to data based on proximity.

(3) A subsystem which uses the indexing structure to evaluate the spatial conditions.

(4) An efficient optimizer to optimize queries that involve both spatial and aspatial selection criteria.

Chapter 3

The Spatial kd-Tree

The kd-tree is well known as an effective data structure for the storage and manipulation of point objects in k-dimensional space [BeF79], however it has been considered unsuitable for non-zero sized objects. In order to extend the kd-tree for non-zero size objects, techniques such as object duplication [MHN84] and object mapping [BaK86] have been proposed. The problems associated with these techniques have been identified in Chapter 2, and lead to performance degradation and extra storage overhead. A new extension of the kd-tree, the **spatial kd-tree** (skd-tree), is proposed in this chapter. This new structure avoids both object duplication and object mapping, and supports two types of search, namely intersection and containment search. Other widely supported structures like R-trees [Gut84] and R^+-trees [SRF87] only directly support intersection search and containment search with these structures is as expensive as intersection search.

This chapter presents the structure of the skd-tree and the operations involved. An empirical analysis of the structure is presented in the next chapter.

3.1 A New-Structure — The Spatial kd-Tree

At each node of a kd-tree, a value (the discriminator value) is chosen in one of the dimensions to partition a k-dimensional space into two subspaces. The two resultant subspaces, *HISON* and *LOSON*, normally have almost the same number of data objects. Point objects are totally included in one of the two resultant subspaces, but non-zero size objects may extend over to the other subspace. Unless we have some means to ensure that while we are examining at one of the subspaces, we do not ignore the objects that extend over from the other subspace, we will have to store the objects that intersect both subspaces at both locations. To avoid the division of objects and the duplication of identifiers, and yet to be able to retrieve all necessary objects, we introduce a virtual subspace for each original subspace such that all objects are totally included in one of the two virtual subspaces. With this method, the placement of an object in a subspace is based solely upon the value of its centroid.

Since a space is always divided into two, we require only one extra value for each subspace: the maximum of the objects in the LOSON subspace, and the minimum of the objects in the HISON subspace, along the dimension defined by the discriminator. Thus,

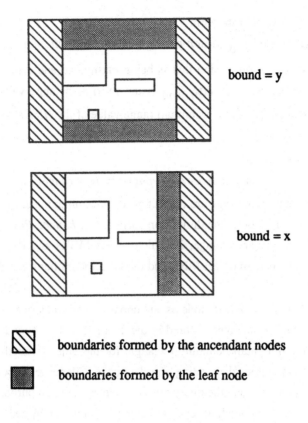

bound = y

bound = x

⬚ boundaries formed by the ancendant nodes

▨ boundaries formed by the leaf node

Figure 3.1 Two possible subspaces of a leaf node

---

the structure of an internal (intermediate) node of the skd-tree consists of

- Two child pointers;
- A discriminator;
- A discriminator-value;
- The maximum value of objects in *LOSON* along the dimension specified by *discriminator*;
- The minimum value of objects in *HISON*.

The maximum range value of *LOSON* is the nearest virtual line that bounds the data objects whose centroids are in the *LOSON* subspace, and the minimum range value of *HISON* is the nearest virtual line that bounds the data objects whose centroids are in the *HISON* subspace.

Hence, internal nodes are of the form

$$(disc, max_{LOSON}, loson\text{-}ptr, disc\text{-}value, hison\text{-}ptr, min_{HISON})$$

where *disc* indicates the dimension that is being partitioned and *disc-value* is the value that partitions the space. The max_{LOSON} is the maximum range value of the *LOSON* subspace and the min_{HISON} is the minimum range value of the *HISON* subspace along the dimension specified by *disc*.

Leaf nodes are of the form

$$[bound, min\text{-}range, page\text{-}ptr, max\text{-}range]$$

where *min-range* and *max-range* are respectively the minimum and maximum values of objects in the data page along the dimension specified by *bound*. *Page-ptr* is the address of a page in the secondary storage, in which object MBRs and identifiers are stored. Throughout this chapter, an object MBR and the object identifier are collectively referred to as a record.

Associated with each leaf node is an unpartitioned subspace which acts as the bounding rectangle for the objects stored in the data page. This bounding rectangle may be larger than the minimum bounding rectangle for the objects stored in the data page, and for intersection queries the reading of a data page is dependent upon the intersection of the bounding rectangle and the query region. Hence, the main purpose of leaf nodes is to reduce its associated bounding rectangle coverage and thereby reduce page accesses. To this end, the *bound* in the leaf node that results in the smallest bounding rectangle is chosen. Figure 3.1 shows two different subspaces are possible depending on the selection of *bound*; in this example, it is clear that the *y* bound yields a smaller resultant subspace.

Figures 3.2a and 3.2b show the structure of a 2-dimensional skd-tree and illustrate the virtual boundary (dotted line), min_{HISON} or max_{LOSON} of each resultant subspace.

Using Pascal-like structure, the leaf node and internal node are defined as follows.

```
type  NODE = record
            disc-value : integer;
            max_LOSON, min_HISON : integer;
            loson-ptr, hison-ptr : integer;
      end;
      LEAF = record
            min-range, max-range : integer;
            page-ptr : integer;
      end;
```

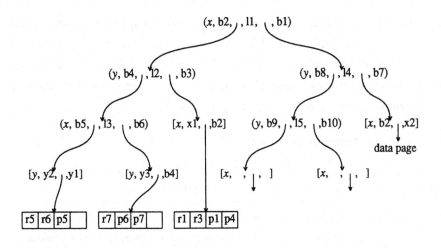

Figure 3.2a The 2-d directory for spatial kd-tree

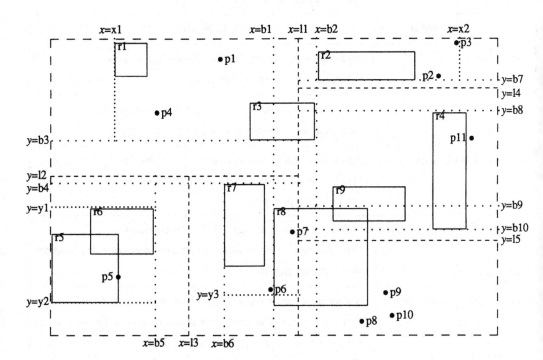

Figure 3.2b The 2-d space coordinate representation

In what follows, we use 0 to indicate the first dimension, and *i-1* to indicate the *i*th dimension of the *k*-dimensional space. We also assume that bounding rectangles are described by arrays of 2*k* elements - br[0..2*k*-1], of which the elements br[2*i*] and br[2*i*+1] define the closed bounded interval along dimension *i*. The macro and functions we use in the algorithms that follow are described in Table 3.1.

Table 3.1 Description of macro/functions

Macro	Description
ORDINATE(br, ty, dim)	*Arguments* — br: bounding rectangle; ty: boolean; dim: dimension; The upper (ty = HI or 1) or lower (ty = LO or 0) ordinate of the bounding rectangle along dimension dim (0 or *x* to indicate X dimension, 1 or *y* Y dimension, ... etc). ORDINATE(br, ty, dim) = br[2*dim + ty].
Functions	**Description**
MAX(x, y)	*Arguments* — x, y: integer or real depending applications. *Value returned* — the maximum value of x and y.
MIN(x, y)	*Arguments* — x, y: integer or real depending applications. *Value returned* — the minimum value of x and y.
PGMAX(pageptr, dim)	*Arguments* — page-ptr: page address; dim: dimension; *Value returned* — the largest ordinate along dimension dim of spatial objects in page page-ptr.
PGMIN(page-ptr, dim)	*Arguments* — page-ptr: page address; dim: dimension; *Value returned* — the smallest ordinate along dimension dim of spatial objects in page page-ptr.
CHECK_LEAF(page-ptr, geo_pred)	*Arguments* — page-ptr: page address; geo_op: spatial predicate; *Purpose* — to read a secondary page and return a set of objects for which the spatial predicate geo_pred is evaluated true. *Values returned* — the set of objects within the page addressed by page_ptr that satisfy the predicate geo_pred.

INTERSECT(br1, br2)	*Arguments* — br1, br2: bounding rectangle; *Purpose* — to check if two bounding rectangles intersect. *Value returned* — *TRUE* if rectangle br1 intersects with rectangle br2 and *FALSE* otherwise. The condition can be formulated as below: ORDINATE(br1, LO, x) ≤ ORDINATE(br2, HI, x) and ORDINATE(br1, LO, y) ≤ ORDINATE(br2, HI, y) and ORDINATE(br1, HI, x) ≥ ORDINATE(br2, LO, x) and ORDINATE(br1, HI, y) ≥ ORDINATE(br2, LO, y).
ADist(obj1, obj2)	*Arguments* — obj1, obj2: spatial objects; *Value returned* — the distance between spatial objects obj1 and obj2 (as discussed in Chapter 2).
Dist(br1, br2)	*Arguments* — br1, br1: bounding rectangles; *Value returned* — the distance between nearest two points of two rectangles; in the case where two rectangles intersect, the distance is zero.
FDist(br1, br2)	*Arguments* — br1, br2: bounding rectangle; *Value returned* — the distance between two points a and b such that a is a point on the boundary of br1 that is nearest to br2, and b is a point on the boundary of br2 furthest from br1.

CUPDATE(S, K, obj)	*Arguments* — S: set of spatial objects; K: integer; obj: spatial object;
	
	Purpose — to find K closest objects to a given object. Note that S will never contain more than K+1 objects.
	
	Value returned — x;
	
	if |S| < K then
	
	        x = ∞;
	
	else
	
	    exclude from S the object that is furthest from obj;
	
	    x = the distance of the furthest object in S from obj;

FUPDATE(S, K, obj)	Arguments — S: set of spatial objects; K: integer; obj: spatial object;
	
	Purpose — to find K furthest objects from a given object. Note that S will never contain more than K+1 objects.
	
	Value returned — x;
	if |S| < K then
	        x = -∞;
	else
	    exclude from S the object that is closest to obj;
	    x = the distance of the closest object in S to obj;

3.2 Searching

In contrast to data structures such as R-trees [Gut84] which support only a single spatial search, namely *intersection search*, skd-trees directly support two types of searches, namely *containment search* and intersection search. Containment search involves the retrieval of data objects which are totally included in a given query region, while intersection search retrieves objects that intersect with the query region. The only difference between these two search algorithms is the area of the space where the intersection testing is done. Containment search allows faster retrieval since it only searches for the points and the centroids of MBRs of non-zero size objects that are contained in the query region. Non-zero size objects are not contained in the query region if their centroids are not included in the query region. The existence of two search strategies does not introduce additional complexity over that of a single search algorithm. On the contrary, it reduces the number of page accesses required and hence improves the retrieval process.

The search algorithm is outlined in a pseudo-code form as follows:

Algorithm Search

Input	node - an intermediate or a leaf node; initially it is the root.
	br - the subspace of the current node; initially the map space.
Output:	A list of objects.
Comment:	Traverse the tree from the root. At each node, if the search is an intersection search, use the virtual boundary; if the search is a containment search, use the tighter boundary.
	geo_pred - spatial predicate.
	search_ty - search type (containment or intersection).
	query_region - the search window declared globally.
	lo_br, hi_lr - bounding rectangles used to define LO and HI subspaces.

SEARCH(node, br)
if node is a leaf node **then**
    ORDINATE(br, LO, node.bound) = node.min-range;
    ORDINATE(br, HI, node.bound) = node.max-range;
    if INTERSECT(br, query_region) **then**
        CHECK_LEAF(node.page-ptr, geo_pred);
    return;
hi_br = br; lo_br = br; /* *rectangles for HI and LO subspaces* */
if search_ty is *containment* search **then**
    ORDINATE(lo_br, HI, node.disc) = MIN(node.disc-value, node.max$_{LOSON}$);
    ORDINATE(hi_br, LO, node.disc) = MAX(node.disc-value, node.min$_{HISON}$);
else /* *an intersection search* */
    ORDINATE(lo_br, HI, node.disc) = node.max$_{LOSON}$;
    ORDINATE(lo_br, LO, node.disc) = node.min$_{HISON}$;
if INTERSECT(lo_br, query_region) **then**
    SEARCH(node.loson-ptr, lo_br);
if INTERSECT(hi_br, query_region) **then**
    SEARCH(node.hison-ptr, hi_br);
end SEARCH.

An implicit rectangular space is associated with each node; the root corresponds to the whole map space, and a leaf node represents an unpartitioned space. The spaces for the nodes appearing on the same level do not overlap, and together they span the rectangular space of parent nodes. These spaces are derived during the traversal of the tree.

Associated with each node are also rectangles which are used during query processing to determine whether the left and right subtrees need to be searched. Depending on the type of the search, containment or intersection, larger or smaller rectangles may be used. In a containment search, the smaller of the maximum range (max_{LOSON}) and *disc-value* is used to determine whether the record may be in the *LOSON*, and the larger value of the minimum range (min_{HISON}) and *disc-value* is used for the *HISON*. In an intersection search, the search space may be larger than the search space in a containment search, hence the min_{HISON} and max_{LOSON} are used instead.

3.3 Insertion

Inserting records for new data objects into an skd-tree is similar to the insertion of points in a point kd-tree. New records are added to a data page, and the page is split if it overflows. As the tree is traversed, the algorithm determines the branching direction of each node and updates the node if the MBR of the object extends over the node boundary. The process of searching is done recursively and on reaching a leaf node, the data page is fetched and insertion may be performed.

Algorithm Insert

Input: node - an intermediate node or a leaf node; initially the root.

Output: the updated skd-tree.

Comment: Use the centroid to determine the place of an object. A split occurs
if the bucket overflows. Variables declared globally:

mbr - the minimum bounding rectangle of the object to be inserted.

C - the centroid of the object to be inserted.

INSERT(node)

if node is a leaf node **then**

    fetch the data page addressed by node.page-ptr;

    if there is room in the data page **then**

        insert the record into the data page;

        for all dimensions (i=0 to k-1) **do**

            get *min-range* and *max-range* along each dimension i;

            calculate the covering rectangle;

        choose the dimension j that has least coverage;

        if j = node.bound **then**

            node.min-range = MIN(ORDINATE(mbr, LO, j), node.min-range);

            node.max-range = MAX(ORDINATE(mbr, HI, j), node.max-range);

        else /* *new boundaries in the leaf node* */

            node.bound = j;

            set new node.max-range and node.min-range;

    else

        SPLIT(node);

        insert the object into the correct data page;

    return;

/* *Internal node* */

if C[node.disc] ≤ node.disc-value **then** /* *the centroid is at the LO subspace* */

    node.max$_{LOSON}$ = MAX(ORDINATE(mbr, HI, node.disc), node.max$_{LOSON}$);

    INSERT(node.loson-ptr);

else /* *the centroid is at the HI subspace* */

    node.min$_{HISON}$ = MIN(ORDINATE(mbr, LO, node.disc), node.min$_{HISON}$);

    INSERT(node.hison-ptr)

end INSERT.

In order to add a new record to a full data page containing M records, it is necessary to divide these $M + 1$ records into two data pages. The division is performed in the dimension which is the largest of the subspace's k-dimensional rectangle. Objects are ordered according to ascending centroid coordinate along the dimension to be partitioned. The objects are divided into two groups, and the line that partitions the rectangle may be anywhere between these two groups. One of the resultant data pages will contain $m + 1$ records, and the other will contain $M - m$ records. The following is the outline of the strategy.

Algorithm Split

Input: node - the leaf node whose data page is to be split.

Output: an internal node pointing to two leaf nodes.

SPLIT(node)

Get the dimension of the longest side of the subspace, let it be i;

Order objects in the data page node.page-ptr in the ascending order of their centroid along dimension i;

Let *disc-value* be halfway between object m and m+1 along dimension i;

Partition the objects with *disc-value* along dimension i;

Create a new leaf node to index the other resultant data page;

Get the *bound*, *min-range* and *max-range* for the two leaf nodes;

Create a non-leaf node to index the two resultant leaf nodes;

Set *disc-value*, and get $\max_{LOSON}$ and $\min_{HISON}$;

end SPLIT.

3.4 Deletion

The deletion of a record will cause a data page to underflow when the resultant number of records is less than half of the page capacity, and this may degrade the storage efficiency. To ensure better storage efficiency the underflowed page is merged with the data page of the sibling leaf node, and the resultant page is re-split in case of overflow. However, this is not possible if the sibling node is not a leaf node. In this case, the following strategy is employed. The algorithm deletes the leaf node and the parent non-leaf node and the pointer which points to the parent node is redirected to the sibling node. As a consequence, the subspace of the sibling node is expanded, and the records in the data page of the deleted leaf node need to be reinserted into the sibling node.

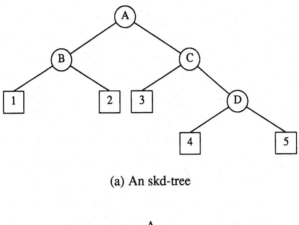

(a) An skd-tree

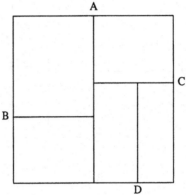

(b) The corresponding planar partition

Figure 3.3 An example for deletion

In the example shown in Figure 3.3a, if the data page indexed by the leaf node 3 has underflowed, node C would be deleted and replaced by node D. Once the records in the deleted data page are reinserted into the subtree of node D, leaf node 3 and the data page can be removed. Figures 3.4a and 3.4b show the resultant tree structure and partitions of subspaces in the case where no overflow occurs after reinsertion. The case where a split occurs after reinsertion is illustrated in Figure 3.5. The method is not only economical with storage, it also incrementally refines the tree structure and hence maintains a more balanced tree as a result.

A rather different reinsertion strategy is employed in the R-tree [Gut84]. In the R-tree, the deletion of a record may cause underflow of a leaf node. In such a case, the entries in the underflowed leaf node must be reinserted into the R-tree from the root.

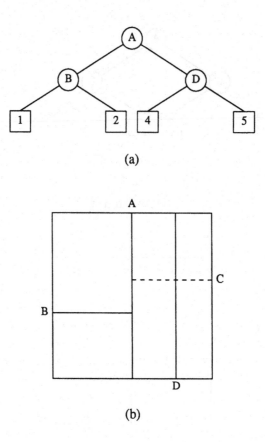

(a)

(b)

Figure 3.4 The expansion of subspaces due to deletion of C

Deleting a leaf node requires the entry in the parent node for that leaf node be deleted and this may cause the underflow as well. If an internal node underflows, all entries must be reinserted from the root to the same level where they resided before the deletion of their node. The reason for reinserting entries from the root is that the covering rectangles of internal nodes may be improved, achieving more efficient coverage. In the skd-tree, the reinsertion of records from the root has the same effect as distributing records in the sibling subtree of the deleted node.

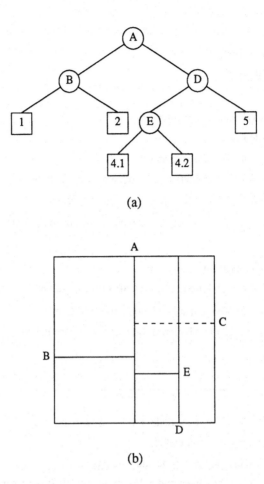

(a)

(b)

Figure 3.5 The deletion of node C and resplit of leaf node 4

The deletion algorithm is outlined as follows.

Algorithm Delete

Input: node - intermediate or leaf; initially the root.

Output: the updated skd-tree.

Comment: Use the centroid of the object to get the page where the object may

reside. Variables declared globally:

mbr - the minimum bounding rectangle of the object to be deleted.

C - the centroid of the object to be deleted.

newbound - the new virtual boundary.

DELETE(node)

if node is a leaf node **then**

    DELETE_OBJ(node);

    return;

if C[node.disc] ≤ node.disc-value **then**

    DELETE(node.loson-ptr);

    /* *update the boundary if necessary* */

    if ORDINATE(mbr, LO, node.disc) = node.max$_{LOSON}$ **then**

        NEWBOUND(node.loson-ptr, node.disc, LOSON)

        node.max$_{LOSON}$ = newbound if the record has been deleted;

else

    DELETE(node.hison-ptr);

    if ORDINATE(mbr, HI, node.disc) = node.min$_{HISON}$ *then*

        NEWBOUND(node.hison-ptr, node.disc, HISON)

        node.min$_{HISON}$ = newbound if the record has been deleted;

end DELETE.

Input:	node - the leaf node whose data page may contain the record intended to be deleted.
Output:	the updated leaf node and data page.
Comment:	Delete the object and redistribute the data if necessary.

DELETE_OBJ(node)

get the data page addressed by node.page-ptr;

delete the record if it is there else return;

if number of record < *m then*

    if the neighboring node is also a leaf node **then**

        merge the two data pages;

        resplit the group into two groups if overflows, as in SPLIT;

    else

        use INSERT to insert the remaining records into the neighboring subtree;

        delete the node and its parent;

        redirect the pointer to the parent, to the neighboring node;

end DELETE_OBJ;

Input: node - an intermediate node or a leaf node.

      i - the dimension where the boundary is changed.

      flag - HISON or LOSON indicator.

Output: a new tighter bound (newbound).

NEWBOUND(node, i, flag)

if node is a leaf node **then**

  if flag = LOSON **then**

    newbound = MAX(newbound, PGMAX(node.page-ptr, i));

  else

    newbound = MIN(newbound, PGMIN(node.page-ptr, i));

  return;

if node.disc $\neq$ i then

  NEWBOUND(node.hison-ptr, i, flag);

  NEWBOUND(node.loson-ptr, i, flag);

else

  if flag = LOSON **then**

    newbound = MAX(node.max$_{LOSON}$, newbound) /* *objects in the data page*

    that is not adjacent to the boundary may contribute the new boundary */

    NEWBOUND(node.hison-ptr, i, flag);

  else

    newbound = MIN(newbound, node.min$_{HISON}$);

    NEWBOUND(node.loson-ptr, i, flag);

end NEWBOUND.

Although the algorithm updates node boundaries, the idea behind the whole strategy is simple. To illustrate an update at work, consider the example shown in Figure 3.6. In this example, the head node may be any intermediate node whose *LOSON* boundary has been affected by a deletion, and readjustment of *node.max$_{LOSON}$* is required, and we shall call this node the *affected node*. Descend the left subtree from the affected node A, and at any level of the subtree, if the node has the same discriminator as that of the affected node (e.g. C, E, F) then only the right subtree is searched, otherwise both subtrees are searched (B, D, G). Only the right subtree in the first case is searched because if there is any object in the left subtree that may contribute to the new boundary, the required value can be detected and obtained from the *max$_{LOSON}$* of the current node, whereas the right bound of the right subtree is unknown. Furthermore, the right subtree contains subspaces that are adjacent to the affected boundary. In the case where a node has a different discriminator from that of the affected node, subspaces of both subtrees of the current node are adjacent to the affected boundary and hence they must be examined. On reaching a leaf node, if a leaf node has the *bound* which is the same as the discriminator of the affected node, then it is only necessary to check the *max-range* value, otherwise the data page has to be read and each record must be checked. The same argument applies to the readjustment of *HISON* boundaries.

Updates of boundaries during deletions may not be required at all if the trade off between efficiency and update cost is not too great. Alternatively, sweeping of a subtree is done after several boundaries in that subtree have been affected.

3.5 Directory Paging

Surprisingly, there is a general misconception that the binary tree is not a suitable data structure for secondary storage based systems. In [BaM72, CeS82, Knu73], however, it has been shown that the binary tree can be paged into secondary storage.

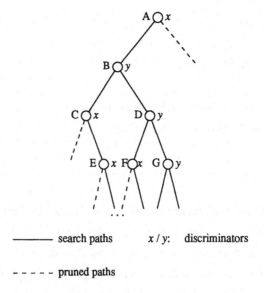

——— search paths x / y: discriminators

– – – – – pruned paths

The boundary of LOSON ($node.max_{LOSON}$) of the first node is affected.

Figure 3.6 Search paths for a tighter boundary

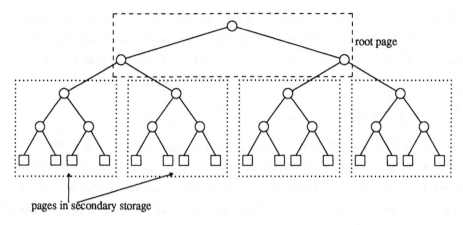

pages in secondary storage

Figure 3.7 Storage structure for the directory

The tree directory can be recursively partitioned into pages as suggested in [Knu73]. A family of internal nodes which is a subtree, is stored in the same page in secondary storage. Grouping the nodes into pages in this form essentially changes the tree from a binary tree to a multiway tree, which has more than two-way branching. In the structure, the page containing the root is usually stored in main memory. Figure 3.7 illustrates the secondary storage structure of an skd-tree, where each secondary page can hold 8 nodes.

3.5.1 Page Structure

Using PASCAL-like syntax, the page structure of the skd-tree is defined as follows:

```
type  PAGE = record
            num-node : integer;
            num-leaf : integer;
            disc : array [1..M] of BIT;
            nodes : array [1..num-node] of NODE;
            leafs : array [1..num-leaf] of LEAF;
      end;
```

The notations used correspond to the definitions of the internal and leaf nodes in Section 3.1. The bit array is used to store the discriminator of each node, and the *num-node* and *num-leaf* are respectively the number of internal and leaf nodes stored in a page. An internal node requires 20 bytes of storage, while a leaf node requires only 12 bytes of storage. For each page, eight bytes are essential to store the counters and an additional $\lceil \frac{M * B}{8} \rceil$ bytes are required for the bit array, where B is the minimum number of bits to sufficiently indicate the direction ($B = 1$ for a 2-dimensional skd-tree and $B = 2$ for a 3 or 4 dimensional skd-tree). The internal nodes are stored sequentially after the parameters (bit array and counters), and then the leaf nodes are stored. Figure 3.8 illustrates a physical page structure, where a page is a continuous block of memory.

3.5.2 Paging Strategy

In [CeS82, p 341], algorithm M is proposed for the paging of AVL tree by Cesarini and Soda. Although we have rather different types of nodes, the approach can be modified for skd-trees. The paging procedure shall be called by the SPLIT routine when a new node and a new leaf node are created. The algorithm used for splitting pages in the skd-tree is outlined as below.

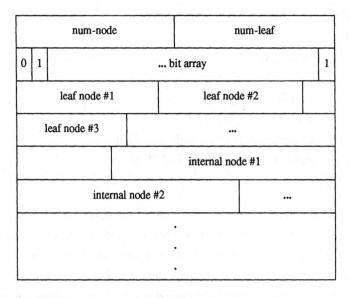

Figure 3.8 The page structure for a spatial 2d-tree

Algorithm Paging

Input: P - the page to be split.

node - an intermediate node to be inserted in page P.

leaf-node - a leaf node to be inserted; this exists only when a node is split.

Output: the updated multiway tree.

PAGE_INSERTION(P, node, leaf-node)

If there is free space in the page P for node and leaf-node **then**

    insert the node, and also leaf-node if it is not NULL;

    update the counters;

    return;

page_root = PAGE_SPLIT(P);

if the parent page of P is not NULL **then**

    PAGE_INSERTION(parent page of P, page_root, NULL);

else

    create new page P';

    PAGE_INSERTION(P', page_root, NULL);

end PAGE_INSERTION.

Input: P - the page to be split.

Output: three updated pages.

PAGE_SPLIT(P)

Create a new page P';

Let the root of the page P be page_root;

Store the LOSON subtree (those nodes residing in P) of page_root in P;

Store the HISON subtree (those nodes residing in P) of page_root in P';

Set all the necessary parameters for P and P';

return(page_root);

end PAGE_SPLIT.

The PAGE_SPLIT procedure that has been presented needs a modification for the following case. When the root node is to be pushed upwards as a result of a page split, the PAGE_SPLIT algorithm assumes that the LOSON and HISON of the root node both reside in the same page. If the root node has only one subtree, however, then no new page should be created and only the root is pushed upwards. In this case, there is no real split. As an example, suppose page 2 in Figure 3.9 has overflowed, then node A is merely pushed upward to page 1 without requiring a split of page 2.

The multiway tree is **not** height-balanced. However, random insertions from a uniform distribution would eventually produce an almost height-balanced multiway tree [Knu73].

3.6 Static Tree Construction

Quite often, an skd-tree is initially built from an available database, in which case the tree can be constructed statically instead of dynamically. Dynamic construction refers to the process of inserting all the records in the database into an initially empty kd-tree by means of the insertion algorithm. Static construction is a preprocessing technique which

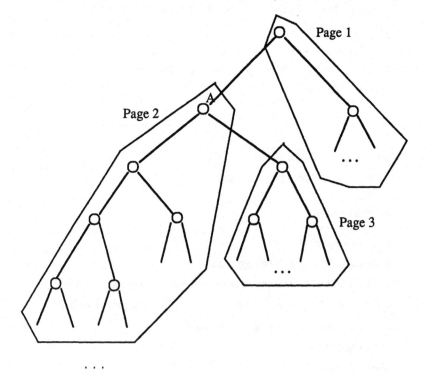

Figure 3.9 A paged skd-tree

packs the tree so that all data pages are as full as possible. Static construction of an skd-tree starts from the root, and then recursively partitions the data objects into two groups. This process is repeated until the size of the groups is less than or equal to M. The algorithm is designed so that close to 100% storage utilization is ensured. With static construction both high storage utilization and an almost balanced tree structures will be obtained. The pseudo code of the algorithm is outlined as follows:

Algorithm Preprocessing

Input: node - an internal node.

           mbr - the subspace that contains all objects.

           list - a list of objects.

Comment: Assume a list of objects available before hand.

           Order the objects (in list) in an ascending order of the centroids along the dimension that the longest side of the minimum bounding rectangle and for the objects then partition the objects into two groups, llist and rlist.

           lmbr, rmbr: bounding rectangles that describe the LO and HI subspaces.

PREPROCESS(node, N, mbr, list)

$N = |\text{list}|$;

$\text{num\_of\_pages} = \lceil \frac{N}{M} \rceil$;

$m_1 = \lfloor \frac{num\ of\ pages}{2} \rfloor \times M$;

$m_2 = N - m_1$;

Use the same method as that of procedure SPLIT to split the N objects into two groups, hlist and llist. Then *HISON* has m_1 objects in hlist and *LOSON* has m_2 objects in llist;

Update lmbr and rmbr;

    /* *lmbr is the left subspace and rmbr is the right subspace.* */

if $m_1 < M$ **then**

        node.hison = a new leaf-node;

        put the objects into a data page;

else

        node.hison = a new internal node;

        PREPROCESS(node.hison, m_2, rmbr, hlist);

/* *end if* */

if $m_2 <$ M **then**

    node.loson = a new leaf-node;

    put the objects into a data page;

else

    node.loson = a new internal node;

    PREPROCESS(node.loson, m_1, lmbr, llist);

set all necessary parameters;

end PREPROCESS.

A statically constructed skd-tree not only lends itself to a better page storage utilization, but is also an almost height-balanced tree. However, the tree may not be perfectly balanced. This is because the preprocessing technique we used clusters the records in such a way that close to 100% utilization of data pages is obtained.

In an environment where insertions and deletions are not frequent once a tree is constructed, we can further pack the directory in the sense that pointers (edges) pointing to the children are not explicitly stored. With this technique, eight bytes of each node are saved, but the disadvantage is that space must be allocated for every possible node at each level. In another words, if the tree is balanced, a great deal of storage can be saved.

A tree is stored as a sequential array, say *arr*, which will be paged into secondary memory. The root of the tree is stored as the first element and its children are stored in *arr[2]* and *arr[3]*. The numbering system follows the breath-first order, from left to right and from top to bottom. An example is shown in Figure 3.10. In general, the left child of node *i* is stored in *arr[2i]*, and the right child of node *i* is stored in *arr[2i + 1]*.

With the sequential storage, leaf nodes and internal nodes utilize the same amount of space. The nodes now have the the following structures:

type NODE = **record**

        disc_value : integer;

        $\max_{LOSON}$, $\min_{HISON}$: integer;

    end;

    LEAF = **record**

        page_ptr : integer;

        min_range, max_range : integer;

    end;

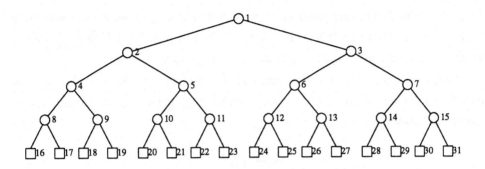

Figure 3.10a Numbering of the nodes for compaction

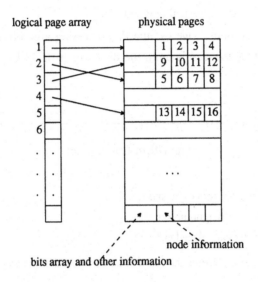

Figure 3.10b The structure of a compacted directory

Consequently, a page has the following format:

```
type snode = record
            int1, int2, int3 : integer;
      end;
      PAGE = record
            bit[1..M] : array of bits;
            arr[1..M] : array of snode;
      end;
```

Notice that in the PAGE structure, there is no bit indicator to distinguish a leaf node from a non-leaf node. However, this is only possible if a data page pointer (*page_ptr*) and a discriminator value (*disc_value*) assume different ranges of values.

Nodes are stored sequentially according to their indexes, and the page number (page#) and the relative location of a node within a page (page_offset) of node *i* can be obtained as follows:

$$page\# = \left\lfloor \frac{i}{M} \right\rfloor,$$

$$page\_offset = (i - 1) \bmod M + 1.$$

The page# is a logical page address, a logical page array (see Figure 3.10b) is used to map this logical address to a physical page number.

To set up such a structure, we use the Preprocessing algorithm given previously to construct the tree. Once the tree is built, the directory is packed by using the following routine.

Algorithm Compaction

Input: root- the root of an skd-tree.

                   ht - the height the skd-tree.

Output: A compact paged skd-tree.

Assumption: The height ht is the depth of a tree where the root has the height of 1.

COMPACT(root, ht)

num_nodes = 2^{ht} - 1; /* *number of node* */

num_pages = num_nodes / M;

for i=1 to num_pages **do**

    get a page, say pg;

    initialize it;

    pageptr[i] = pg;

PLACENODE(root, 1);

end COMPACT.

Input: node - An intermediate or a leaf node.

      nodenumber - the representative number of the current node.

Output: An updated page.

PLACENODE(node, nodenumber)

page_num = $\lceil$ nodenumber / M $\rceil$;

off_set = (nodenumber - 1) mod M + 1;

page_ptr = pageptr[page_num];

store node in page page_ptr at the relative location off_set;

set all necessary parameters;

if node is a leaf-node **then**

  return;

PLACENODE(node.loson-ptr, nodenumber*2);

PLACENODE(node.hison-ptr, nodenumber*2 + 1);

end PLACENODE.

3.7 Data Storage Structure

It is obvious that spatial description of objects are of variable length. One proposed approach to the storage of the spatial description is to use a conventional relation [AbS86]. Alternatively, the spatial data can be stored in an external file, and accessed using a *geographic object identifier* (*gid*), a unique identifier for each spatial object. A *gid* can be a logical identifier (surrogate) or a physical address. In our implementation, we took the latter approach to make query processing more efficient. The *gid* is structured as the concatenation of *page#* and *offset_index*, where the *page#* refers to the page on which the object is stored and the *offset_index* provides the location of a pointer to the object that is stored in the page. With this approach, a spatial object can be moved within a physical page without changing its *gid*. Figure 3.11 illustrates the global data storage arrangement for the aspatial data, spatial index, and the spatial data. Notice that the *gid* is stored as a relation attribute value, and also as an entry in the data pages of skd-trees. The *object data file* contains the complete spatial definitions of the geographic objects and therefore the spatial information can be obtained in one disk access once the *gid* is known. With this organization, the spatial and aspatial information are stored separately. Efficient access to aspatial data is supported and an integrated access to both spatial and aspatial informations is maintained.

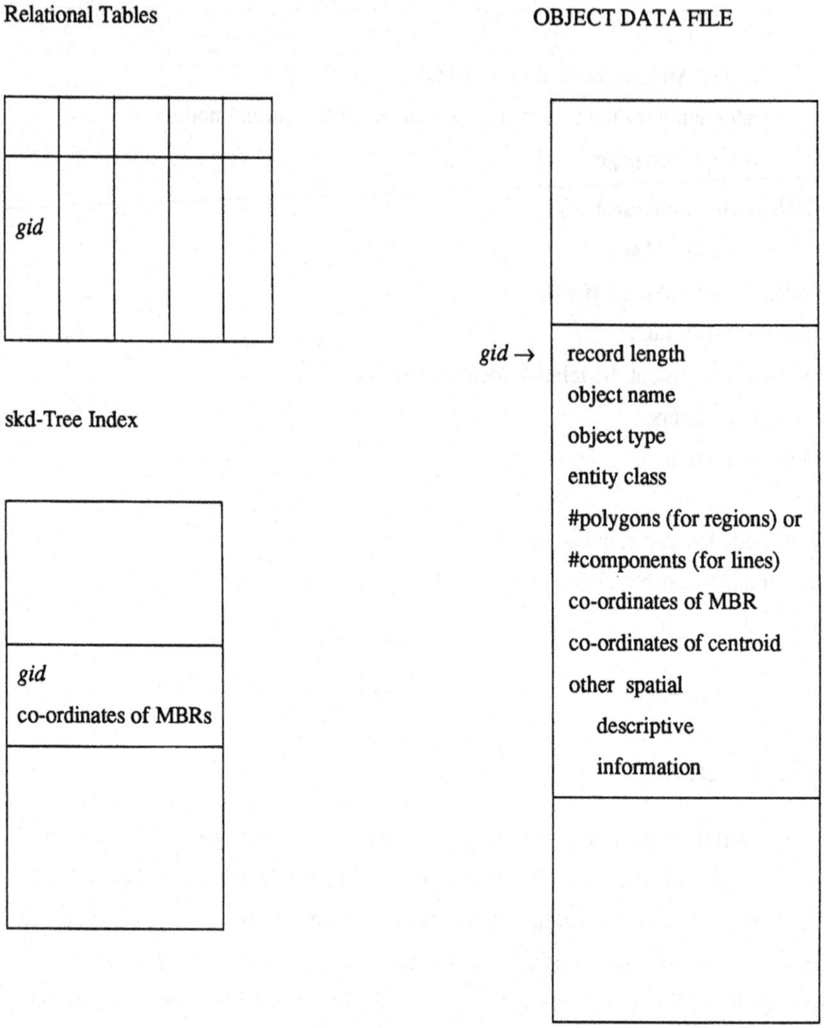

Relational Tables   OBJECT DATA FILE

Figure 3.11 Data storage structure for a GIS

   To illustrate how the structure may be used, consider Example 3.1. By restricting the relation *road* on the attribute *name*, we get a resultant temporary relation for which all values of road.name are 'Wellington'. Since *gid* is one of the attributes of this temporary relation, we can obtain the spatial description of all roads appearing in the resultant relation from the object data file for display purposes.

```
Example 3.1. Find all roads whose name is 'Wellington'.
Select    *
From      road
Where     road.name = 'Wellington'.
```

Example 3.2 illustrates an instance where the use of skd-trees may be required.

```
Example 3.2. Find all railways that intersect with
          'Wellington' road.
```

```
Select    railway.name
From      road, railway
Where     road.name = 'Wellington' and
          road intersects railway.
```

One way of processing the query is to firstly restrict relation *road*, and then to use each object of the resultant temporary relation to search the skd-tree for all railways that intersect with the given road.

3.8 Spatial Primitive Operations

As mentioned in Chapter 2, most GISs differ substantially on the operations they support. Although GEOQL does not attempt to cover all possible operations, it supports most of the fundamental operators. In this section, the definition of each operator supported by GEOQL [SMO87] is given, and the approach to implementing each of these operators is described.

Definition 3.1. Object A *intersects* object B if and only if these two objects share a common point.

Definition 3.2. Object A is *adjacent* to object B if and only if they intersect, and every point of intersection is a boundary point of both objects.

Definition 3.3. Object A *contains* object B if and only if every point of B is a point of A.

Definition 3.4. Object A *ends at* object B if and only if they intersect and every point of intersection is an extreme point of A.

Definition 3.5. Object A is *situated at* object B if and only if every point of A is a point of B and vice-versa.

Definition 3.6. Object A *joins* object B if and only if the intersection points are extreme points.

Definition 3.7. Object B exists *within* X units of object B if and only if every point of B is a point of the extended virtual object of A, which is obtained by extending out the boundary of object A by X.

Definition 3.8. Object B is *closest* to object A if and only if, for all objects that are of interest, no other object is closer to A than B.

Definition 3.9. Object B is *furthest* from object A if and only if, for all objects that are of interest, no other object is further from A than B.

The fundamental spatial operations are intersection and containment; they form the basis of other spatial operations. Unlike other spatial tree structures, skd-trees support both containment and intersection searches. For all operations, except closest and furthest, two basic searches are used to obtain the date pages, and then a particular spatial testing may be performed. While predicates *intersect, adjacent, end_at,* and *joins* use intersection search to search the tree, predicates *situated_at, within* and *contains* take advantage of the more efficient containment search to get the records that may qualify.

Unfortunately, operators involving measurement of distance (*closest* and *furthest*) cannot use the existing search routines. In fact, they are the most expensive of all operators to support and the corresponding searches require retrieving a large proportion of the database. To reduce the search space, a heuristic can be incorporated in the search strategies to prune the search space. There are various measures of distance between two irregularly shaped objects that can be supported. Two measurements often used are:

(1) the distance between the two nearest points of the two objects,
(2) the distance between the centroids of the two objects.

Another measure used is the distance between the MBRs of the two objects. This measure can be used as a filter to dynamically reject those objects that are obviously not candidates for a query based on measures (1) or (2). In the following algorithms, function Dist returns the distance between the nearest two points of two objects; in the case where two objects intersect, the distance is zero. The algorithm Closest is outlined below.

Algorithm Closest

Input:	obj - the given object.
	node - an intermediate or a leaf node; initially the root.
	br - the subspace of the current node.
Output:	the closest object to object obj.
Comment:	lo_br, hi_br - bounding rectangle to describe LO and HI subspaces.
	X - ∞ initially. X stores the current estimate of the distance between obj and the closest database object.

CLOSEST(obj, node, br)
if node is a leaf-node **then**
    ORDINATE(br, LO, node.bound) = node.min-range;
    ORDINATE(br, HI, node.bound) = node.max-range;
    if Dist(br, obj.mbr) < X **then**
        for each record e of the data page indexed by node.page-ptr **do**
            if Dist(e.mbr, obj.mbr) < X **then**
                fetch the actual object of e;
                check the object and update X if necessary;
    return;
lo_br = br; hi_br = br;
ORDINATE(lo_br, HI, node.disc) = node.max$_{LOSON}$;
ORDINATE(hi_br, LO, node.disc) = node.min$_{HISON}$;
if Dist(lo_br, obj.mbr) < X **then**
    CLOSEST(obj, node.loson-ptr, lo_br);
if Dist(hi_br, obj.mbr) < X **then**
    CLOSEST(obj, node.hison-ptr, hi_br);
end CLOSEST.

In the Furthest algorithm, a different distance measuring criterion is required. Even if a bounding rectangle, *mbr*, intersects with an object, *obj*, one of *mbr*'s objects may be the furthest object from *obj*. For this reason we introduce FDist, a measure, which returns the distance between two points *a* and *b* such that *a* is a point on the object *obj* nearest to the bounding rectangle, and *b* is a point on the bounding rectangle furthest from the object.

Algorithm Furthest

Input: obj - the given object.

                 node - an intermediate or a leaf node; initially the root.

                 br - the subspace of the current node.

Output: the furthest object from obj.

Comment: lo_br, hi_br - bounding rectangles to describe LO and HI subspaces.

                 X - -∞ initially. X represents the current estimate of the maximum

                 distance between obj and database object.

FURTHEST(obj, node, br)

if node is a leaf-node **then**

    ORDINATE(br, LO, node.bound) = node.min-range;

    ORDINATE(br, HI, node.bound) = node.max-range;

    if FDist(br, obj.mbr) > X **then**

        for each record e of the data page indexed by node.page-ptr **do**

            if Dist(e.mbr, obj.mbr) > X **then**

                fetch the actual object of e;

                check the object and update X if necessary;

    return;

lo_br = br; hi_br = br;

ORDINATE(lo_br, HI, node.disc) = node.max$_{LOSON}$;

ORDINATE(hi_br, LO, node.disc) = node.min$_{HISON}$;

if FDist(lo_br, obj.mbr) > X **then**

    FURTHEST(obj, node.loson-ptr, lo_br);

if FDist(hi_br, obj.mbr) > X **then**

    FURTHEST(obj, node.hison-ptr, hi_br);

end FURTHEST.

3.9 Set Operations

Most GEOQL spatial predicates are binary predicates. To implement such a GEOQL predicate, a loop iteration is required. For example, to evaluate a spatial predicate such as road intersects railway assuming that railway is indexed, the following strategy is possible.

```
for each object in road do
     search the skd-tree of railway and
     evaluate the predicate.
```

The point we are trying to make here is that since a segment of relation road is read at a time, it would be handy to search the skd-tree with a set of railways instead of with a single railway. A substantial cost would be saved if the set used is large. Let the set of objects be O, then the searching strategy can be modified as below:

Algorithm Set_Search

Input:	node - an intermediate or a leaf node; initially it is the root.
	br - the subspace of the current node; initially the map space.
	O - a set of query regions.
	search_ty - search type (intersection or containment).
Output:	A list of objects.

SET_SEARCH(node, br, O)
if node is a leaf node **then**
    ORDINATE(br, LO, node.bound) = node.min-range;
    ORDINATE(br, HI, node.bound) = node.max-range;
    if $\exists$ o $\in$ O s.t. INTERSECT(br, o) **then**
        CHECK_LEAF(node.page-ptr);
    return;
hi_br = br; lo_br = br;
if search_ty is *containment* **then**
    ORDINATE(lo_br, HI, node.disc) = MIN(node.disc-value, node.max$_{LOSON}$);
    ORDINATE(hi_br, LO, node.disc) = MAX(node.disc-value, node.min$_{HISON}$);
else
    ORDINATE(lo_br, HI, node.disc) = node.max$_{LOSON}$;
    ORDINATE(hi_br, LO, node.disc) = node.min$_{HISON}$;

$O_{LOSON} = \emptyset; \ O_{HISON} = \emptyset;$

for each $o \in O$ **do**

    if INTERSECT(lo_br, o) **then**

        $O_{LOSON} = O_{LOSON} \cup o;$

    if INTERSECT(hi_br, o) **then**

        $O_{HISON} = O_{HISON} \cup o;$

if $O_{LOSON} \neq \emptyset$ **then**

    SET_SEARCH(node.loson-ptr, lo_br, O_{LOSON});

if $O_{HISON} \neq \emptyset$ **then**

    SET_SEARCH(node.hison-ptr, hi_br, O_{HISON});

end SET_SEARCH.

At an internal node, a set of query regions that intersect with the present covering subspace is used to determine a subset of the query regions for the subtree traversal. The tree is searched recursively with a set of query regions; the set may get smaller as the depth of the tree increases. A similar approach can be used for INSERT or DELETE.

Set searches are not applicable to *closest* and *furthest* searches. However, instead of searching for a single object, an skd-tree may be searched for a set of objects. The following two algorithms return a set of objects that are closest to and furthest from a given object, respectively. The algorithms use the functions CUPDATE and FUPDATE in Table 3.1 for updating the current set of K closest objects and K furthest objects respectively.

Algorithm Set_Closest

Input: obj - the given object.

node - an intermediate or a leaf node; initially the root.

br - the subspace of the current node.

Output: a set of K closest objects to object obj.

Comment: Similar to the Closest algorithm except that a set of closest objects

is maintained. Variables used:

S - a set of objects that have been retrieved so far; initially empty;

K - the number of closest objects to be retrieved;

$X - \infty$ initially;

lo_br, hi_br - bounding rectangles to describe LO and Hi subspaces.

SET_CLOSEST(obj, node, br)

if node is a leaf node **then**

ORDINATE(br, LO, node.bound) = node.min-range;

ORDINATE(br, HI, node.bound) = node.max-range;

if Dist(br, obj.mbr) < X **then**

for each record r of the data page indexed by node.page-ptr **do**

if Dist(r.mbr, obj.mbr) < X **then**

fetch the actual object of r;

d = ADist(r, obj);

if (d < X) **then**

$S = S \cup \{r\}$;

X = CUPDATE(S, K, obj);

return;

lo_br = br; hi_br = br;

ORDINATE(lo_br, HI, node.disc) = $node.max_{LOSON}$;

ORDINATE(hi_br, LO, node.disc) = $node.min_{HISON}$;

if Dist(lo_br, obj.mbr) < X **then**

SET_CLOSEST(obj, node.loson-ptr, lo_br);

if Dist(hi_br, obj.mbr) < X **then**

SET_CLOSEST(obj, node.hison-ptr, hi_br);

end SET_CLOSEST.

Algorithm Set_Furthest

Input: obj - the given object.

node - an intermediate or a leaf node; initially the root.

br - the subspace of the current node.

Output: the K closest objects from obj.

Comment: Variables used:

S - a set of objects that have been retrieved so far; initially empty.

K - the number of closest objects to be retrieved.

X - $-\infty$ initially.

lo_br, hi_br - bounding rectangles to describe LO and HI subspaces.

SET_FURTHEST(obj, node, br)

if node is a leaf node **then**

  ORDINATE(br, LO, node.bound) = node.min-range;

  ORDINATE(br, HI, node.bound) = node.max-range;

  if FDist(br, obj.mbr) > X **then**

      for each record r of the data page indexed by node.page-ptr **do**

        if Dist(r.mbr, obj.mbr) > X **then**

          fetch the actual object of r;

          d = ADist(r, obj);

        if (d > X) **then**

          S = S $\cup$ {r};

          X = FUPDATE(S, K, obj);

    return;

lo_br = br; hi_br = br;

ORDINATE(lo_br, HI, node.disc) = node.max$_{LOSON}$;

ORDINATE(hi_br, LO, node.disc) = node.min$_{HISON}$;

if FDist(lo_br, obj.mbr) > X **then**

  SET_FURTHEST(obj, node.loson-ptr, lo_br);

if FDist(hi_br, obj.mbr) > X **then**

  SET_FURTHEST(obj, node.hison-ptr, hi_br);

end SET_FURTHEST.

In the algorithm SET_CLOSEST, a parameter X (initially a large number) is used to compare the distance between an object and the query object. If the distance is less than X, then the object is included in the set S. When S contains K objects, X represents the distance between the furthest object in S from the query object. If a new object is included in S, then X will be updated. A similar approach is adopted for the SET_FURTHEST algorithm.

3.10 Summary

Although kd-trees can be used to efficiently index point data, none of the previous proposed extensions, such as 4d-trees [BaK86] or mkd-trees [MHN84], have proved to be an efficient spatial structure for non-zero sized objects. The major contribution of this chapter is the proposal of a new kd-tree structure called the skd-tree. The algorithms for inserting, updating and searching the skd-tree have been presented. The skd-tree provides two types of search, one of which is not directly supported by other existing structures. To do a containment search using for for example R-trees [Gut84] or R$^+$-trees [SRF87], an intersection search is performed on the directory and containment testing is performed only on the leaf nodes.

Chapter 4

Performance Analysis and Case Studies

In a database system, the auxiliary data structures are used to index the data for efficient query processing. In order to show that the skd-tree is an efficient indexing structure, we conducted several simulations and the results are reported in this chapter.

As has been explained in Chapter 3 the leaf nodes may be removed from the skd-tree in order to save storage space. Leaf nodes serve as a filter to reduce data page reads. The number of leaf nodes is about half of the total number of nodes, and they occupy about one-third of the storage of the directory. Are leaf nodes effective? In order to answer this question, we conducted some experiments on the skd-tree with only a single type of node.

4.1 Empirical Analysis

In Chapter 2, we surveyed different approaches taken to index spatial objects and compared them intuitively. Among the structures discussed in Chapter 2, the R-tree proposed in [Gut84] is widely recognized as an efficient spatial indexing structure [RoL85, SRF87, Sto86]. It is an ideal indexing structure for comparison with the skd-tree.

4.1.1 Quantities of Interest

Two common criteria used in selecting an efficient indexing structure are the *storage requirement* and the *query efficiency*.

The storage requirement is the number of pages required to store both the indexes and the data. Note that since indexing structures duplicate some data objects, we include in the storage requirement the storage overheads for the data as well as the index.

The *query cost* of a search is defined to be the number of I/O operations required (the number of pages accessed), and hence the query efficiency is an average search cost for a set of given queries (total cost for a set of queries divided by the number of queries).

A more complicated query cost includes the computation cost. However, since page accesses are the dominant cost and the computation costs are affected by other factors such as programming styles and programming languages, it is acceptable to use a cost measure based on page accesses.

4.1.2 Parameters

The parameters of the experiments are the page sizes, which determine the number of nodes M and data objects that can be stored in a page. The page sizes used are 128, 256, 512, 1024, and 2048 bytes.

The density, D, of a single point location, is the number of spatial objects that contain that point. For a map space, we define the average density as the sum of the areas of the objects divided by the area of the map space, i.e.

$$D = \frac{(\sum_{i=1}^{N} \text{area of object}_i)}{\text{area of map space}}$$

where N is the number of objects. Thus the density is greatly affected by the population and size of spatial objects. The density provides an indication of the number of rectangles that overlap on average.

4.1.3 Simulation Methodology

In the first experiment [OMS87a], we compare the skd-tree with the R-tree.

The leaf nodes of R-trees play the same role as data pages of skd-trees, and are used for the storage of objects (MBRs and identifiers). When presenting the result, the references to data pages refer to the leaf nodes of R-trees and the data pages of skd-trees. In the experiments, we have assumed that there are very few points (less than M) or centroids of MBRs that have the same point locations. If this is not true, then the split algorithm in the skd-tree needs to split a data page into two groups such that one of the groups accommodates the objects that have the same point locations.

In the R-tree, the entries require 20 bytes of storage and a node requires extra 8 bytes to store the type of the node (leaf or non-leaf) and the number of entries in that node. The maximum number of entries in an R-tree node is determined by the page size.

In this experiment, 10,000 2-dimensional data objects, points and non-zero sized MBRs, and 500 query regions were used. All these data objects were randomly generated within the given ranges, $0 \leq x \leq 5200$, $0 \leq y \leq 5200$. The MBRs and query rectangles are aligned with the two conventional axes, X and Y. The extent of MBRs are bounded by 200; this gives a density of close to 3. Th extent of query rectangles are between the range of 0 to 200 and the query rectangles are used in both containment search and intersection search queries of the following forms:

Containment search: *Find all data objects that are contained in a given query rectangle.*

Intersection search: *Find all data objects that intersect with a given query rectangle.*

The skd-trees were constructed both dynamically and statically, with the directory stored as pages. Containment and intersection searches were performed on the spatial kd-tree.

The R-tree was constructed dynamically and only intersection search was performed. For the R-tree, there is no distinction between containment and intersection testing during the traversal of the internal nodes. The containment test may only be performed on the MBRs of the real objects, which reside in the leaf-nodes.

Tables 4.1a, 4.1b and 4.1c summarize the performance of both spatial kd-tree and R-tree. The data pages refers to the leaf nodes of R-trees and the data pages of skd-trees. The depth is the depth of the directory, excluding the data pages. It is assumed that the root pages of both skd-trees and R-trees are stored in secondary storage.

The empirical results show that the skd-tree requires more storage than the R-tree, but the skd-tree reduces search time significantly, with savings of up to 40% in disk accesses. The results confirm the prediction of Chapter 3, that the containment search of the skd-tree is more efficient than the intersection search. Hence, it is advantageous to have two types of searches.

4.2 A Variant of the skd-Tree

In the skd-tree, the leaf nodes make up about half the number of nodes. One-third of the storage can be saved if the leaf nodes are removed. The role of the leaf node is to reduce the number of the data page accesses. In this section, we investigate the effectiveness of the leaf nodes by performing experiments with skd-trees that have only a single type of node. The extension is illustrated in Figure 4.1. To distinguish this extension from the conventional skd-tree, we call it the homogeneous skd-tree — hskd-tree. We use the same data and page sizes as in Section 4.1. The results are presented in Tables 4.2a and 4.2b. The partitioning and the paging remain the same, and the number of data pages remains unchanged; therefore they are not stated in the tables.

The results show that the leaf nodes are required to reduce data page accesses. Note that the intersection search of the hskd-tree is as efficient as that of the R-tree. Yet the hskd-tree provides more efficient containment search.

Table 4.1a R-tree experimental results

Page size	Depth	Storage(pages)		Query Efficiency
		Directory	Data	Intersection
128	5	683	2318	23.2
256	4	155	1161	14.8
512	3	35	561	8.0
1024	2	9	285	5.8
2048	2	3	142	4.2

Table 4.1b Static spatial kd-tree experimental results

Page size	Depth	Storage(pages)		Query Efficiency	
		Directory	Data	Intersection	Containment
128	5	606	1709	6.5	4.4
256	3	141	863	5.5	4.1
512	2	36	413	3.9	3.1
1024	2	11	212	3.3	2.8
2048	2	3	107	2.9	2.6

Table 4.1c Dynamic spatial kd-tree experimental results

Page size	Depth	Storage(pages)		Query Efficiency	
		Directory	Data	Intersection	Containment
128	6	1004	2315	14.9	9.5
256	4	232	1181	8.5	5.9
512	3	55	567	5.6	4.3
1024	2	11	275	3.6	2.9
2048	2	5	143	3.1	2.7

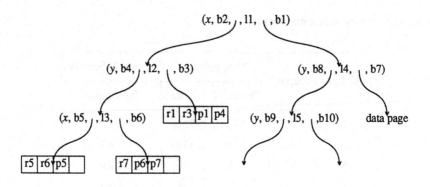

Figure 4.1a The 2-d directory for a variant of the skd-tree

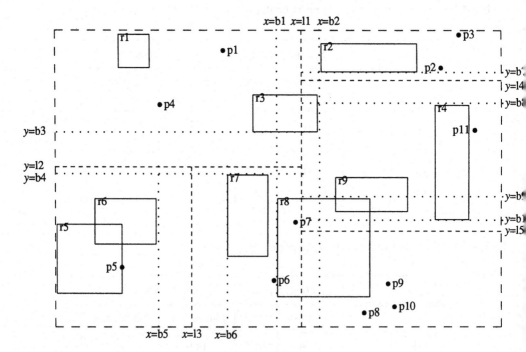

Figure 4.1b The 2-d space coordinate representation

Table 4.2a Static hskd-trees experimental results

Page size	Depth	Directory	Query Efficiency	
		Storage	Intersection	Containment
128	5	368	9.4	15.2
256	3	104	5.9	9.3
512	2	22	4.0	5.9
1024	2	9	3.7	5.0
2048	2	3	3.4	4.2

Table 4.2b Dynamic hskd-trees experimental results

Page size	Depth	Directory	Query Efficiency	
		Storage	Containment	Intersection
128	4	575	10.6	17.1
256	4	143	7.1	11.2
512	3	34	5.3	7.7
1024	2	9	3.8	5.1
2048	2	3	3.5	4.4

4.3 Evaluation with Non-uniform Data

Many economical and physiological phenomena obey Zipf's law [Zip49], or the 80/20 rule [Hei63]. Such distributions are *highly skewed*, and therefore, definitely **not** uniform. It is not suggested here that most of the distribution of spatial objects and their sizes *exactly* follow Zipf's law, but rather, they are skewed. For instance, in a particular space, it is most likely that the majority of the rectangles are small and only a minority are large. In this context, the measurement of one rectangle is relative to a general class of the rectangles. Another instance of skewedness is the location of the rectangles; certain locations are likely to be denser (more rectangles covering the same area) than the others. In general, the distribution of the data can either follow one or both of the following skew conditions:

(1) The sizes of the objects are skewed; smaller objects occur more frequently than larger objects.

(2) The locations of objects are skewed, i.e., the densities are skewed. This is very common in GISs, where a large number of objects occupy certain neighbourhoods and a smaller number of objects are scattered over a relatively large area.

4.3.1 The Skew Factor

Zipf's law is defined as

$$p_1 = \frac{c}{1}, \quad p_2 = \frac{c}{2}, ..., \quad p_N = \frac{c}{N} \text{ ,where } c = \frac{1}{H_N} \text{ and } H_N = \sum_{i=0}^{N} \frac{1}{i}$$

where $p_1 + p_2 + ... + p_N = 1$ and $p_1 \geq p_2 \geq ... \geq p_N$, and N is the number of different events.

A simpler distribution function using the approximation that

$$n^\theta - (n-1)^\theta \approx \theta n^{(\theta-1)}(1+O(1/n))$$

is given by [Knu73] as follows:

$$p_1 = \frac{c}{1^{(1-\theta)}}, \quad p_2 = \frac{c}{2^{(1-\theta)}}, ..., \quad p_N = \frac{c}{N^{(1-\theta)}} \text{ ,where } c = \frac{1}{H_n^{(1-\theta)}}$$

$H_N^{(1-\theta)}$ is the Nth harmonic number to the order of $(1-\theta)$ and is given by:

$$H_N^{(1-\theta)} = \sum_{i=1}^{N} \frac{1}{i^{(1-\theta)}}$$

We define θ as the *skew factor* where $\theta \leq 1$. A smaller θ indicate a more skewed data. The important properties of the above approximation is when

- $\theta = 1$ the probability distribution approximates a uniform distribution;
- $\theta = \log 0.8/\log 0.2$ the probability distribution approximately follows 80/20 rule;
- $\theta = 0$ the probability distribution approximate a Zipf's distribution of order 1.

4.3.2 Data Sets

Three sets of 12000 rectangles were generated with skew factor θs being 0, 0.5 and 1. The width (along X direction) and the height (along Y direction) of rectangles are bounded by the interval [1, 3000]. The N (number of classes) is 100, and the first class of lengths is within the range of [1, 30], the second class [31, 60], ..., and the Nth class [2971, 3000]. The Zipf's distribution specifies that the first class objects occur most often and the last one least frequent. Thus when θ is zero, the number of smaller rectangles is much larger than the number of bigger rectangles. When θ is one, the distribution is uniform.

Using these three sets of rectangles, different maps (distribution of rectangles on a global space) may be formed. Once again, three different θ values are used: 0, 0.5 and 1. The global space is defined by (0,0) and (100000, 100000). Both sides of the space are partioned into equal lengths with interval of 500, i.e. a space is partitioned into smaller 500 × 500 equal sized cells. The intervals are randomly ranked such that the Zipf's function will allocate more rectangles to certain cells than the others. When the placements and the rectangle sizes are uniform, the density is about 3. With θ value being 0, certain locations may be very dense.

In total, we have 3 sets of objects × 3 map distributions = 9 data sets.

4.3.3 Empirical Results

As can be observed from the previous results, the skd-tree requires more storage than the R-tree for the directory. To reduce the storage requirement, a node page of an skd-tree is allowed to store more than one subtree. Further instead of storing the values of max_{LOSON} and min_{HISON}, the offsets (2 bytes each) of max_{LOSON} and min_{HISON} from the partitioning line *disc_value* are stored. With this compaction, less storage is required.

In this experiment, the packed R-tree algorithm [RoL84, RoL85] is used to construct the R-tree statically; the tree is built using preprocessing. As pointed out in Chapter 2, the nearest function is not defined in [RoL84, RoL85], we use the distance between the centroids of MBRs to determine the nearest objects. During the search for the next nearest object to be included into the current node, a pruning strategy is employed to reduce the search time. It should be noted that the pruning does not affect the packing algorithm given in [RoL84].

With 9 sets of data, 4 indexing structures (static and dynamic of R-trees and skd-trees), and 5 different page sizes, we have a total of 180 (9 × 4 × 5) experiments. Results are grouped with respect to the page size.

Two depths of skd-trees are shown in the tables: the depth in terms of the number of pages and the depth of the binary tree in terms of the skd-tree's nodes (this is the figure given in the bracket).

In the tables where the pages accesses are given, the values are the number of pages (directory and data pages) read for a total of 10,000 queries. Each row contains the results for a particular distribution of object sizes. Each column gives results for the same distribution of object locations. Results are given both for the dynamic tree construction and for the static tree construction. The static R-tree is referred to as the packed R-tree [RoL84, RoL85].

Table 4.3 Empirical results for page size = 2048 bytes

Page Accesses for 10000 Queries							
Structures	$\theta_{object\ size}$	$\theta_{location}=0$		$\theta_{location}=0.5$		$\theta_{location}=1$	
		static	dynamic	static	dynamic	static	dynamic
R-tree	0	41132	44699	43456	41106	44553	40218
	0.5	42287	48156	45693	41666	47564	42675
	1	43167	48009	46682	46165	48187	42483
skd-tree intersection search	0	32864	36754	32806	35350	32160	34150
	0.5	34182	38321	33874	37106	33629	36182
	1	34863	39048	34549	37857	34383	37122
skd-tree containment search	0	28967	32035	28796	30752	28439	29987
	0.5	29057	32154	28764	30746	28460	29993
	1	29090	32177	28762	30777	28465	30012

Storage Usage and Description				
Structures	$\theta_{object\ size}$	Depth	Directory Pages	Data Pages
Packed R-tree	0	2	3	119
	0.5	2	3	119
	1	2	3	119
R-tree	0	2	3	161-176
	0.5	2	3	160-171
	1	3	2	161-167
Static skd-tree	0	2(7)	2	118-122
	0.5	2(7)	2	118-122
	1	2(7)	2	118-122
Dynamic skd-tree	0	2(7)	3	166-171
	0.5	2(7)	3	166-173
	1	2(7)	3	167-173

Table 4.4 Empirical results for page size = 1024 bytes

Page Accesses for 10000 Queries							
Structures	$\theta_{object\ size}$	$\theta_{location}=0$		$\theta_{location}=0.5$		$\theta_{location}=1$	
		static	dynamic	static	dynamic	static	dynamic
R-tree	0	49052	52348	48718	52231	51469	49914
	0.5	48582	52979	51919	53812	52127	53389
	1	51531	55609	53792	53138	55364	55434
skd-tree intersection search	0	39459	42393	40320	42537	40225	43184
	0.5	41333	45514	42596	45256	42635	46442
	1	42393	46232	43824	46587	43918	48074
skd-tree containment search	0	33492	34823	34103	35568	34564	36158
	0.5	33593	35614	34178	35738	34633	36302
	1	33648	35135	34204	35516	34648	36209

Storage Usage and Description				
Structures	$\theta_{object\ size}$	Depth	Directory Pages	Data Pages
Packed R-tree	0	2	6	241
	0.5	2	6	241
	1	2	6	241
R-tree	0	2	10-11	346-359
	0.5	2	9-12	336-348
	1	2	9-11	344-340
Static skd-tree	0	2(8)	8	240-242
	0.5	2(8)	8	240-242
	1	2(8)	8	240-242
Dynamic skd-tree	0	2(9)	12-13	331-340
	0.5	2(9)	12-13	336-342
	1	2(9)	12-13	336-338

Table 4.5 Empirical results for page size = 512 bytes

Page Accesses for 10000 Queries							
Structures	$\theta_{object\ size}$	$\theta_{location}=0$		$\theta_{location}=0.5$		$\theta_{location}=1$	
		static	dynamic	static	dynamic	static	dynamic
R-tree	0	51457	70147	53772	82596	56183	79494
	0.5	53186	69668	58589	87188	59972	76682
	1	56683	79266	61649	88705	62948	76383
skd-tree intersection search	0	47329	61254	47618	60919	48353	62071
	0.5	51033	66139	51945	66587	52673	67623
	1	52953	68506	54298	69660	55154	70705
skd-tree containment search	0	37611	49193	38771	50209	39577	51251
	0.5	37788	49620	39018	50478	39741	51642
	1	37917	49693	39088	50608	39799	51688

Storage Usage and Description				
Structures	$\theta_{object\ size}$	Depth	Directory Pages	Data Pages
Packed R-tree	0	2	21	481
	0.5	2	21	481
	1	2	21	481
R-tree	0	3	40-45	674-687
	0.5	3	41-42	689-692
	1	3	37-42	687-688
Static skd-tree	0	2(10-11)	33-34	481-487
	0.5	2(10-11)	33	481-487
	1	2(10-11)	33	481-487
Dynamic skd-tree	0	3(10-11)	47-49	687-693
	0.5	3(10-11)	48-50	686-699
	1	3(10)	47-49	691-695

Table 4.6 Empirical results for page size = 256 bytes

Page Accesses for 10000 Queries							
Structures	$\theta_{object\ size}$	$\theta_{location}=0$		$\theta_{location}=0.5$		$\theta_{location}=1$	
		static	dynamic	static	dynamic	static	dynamic
R-tree	0	72733	119541	78718	127920	79073	116656
	0.5	78587	102734	83167	188772	90157	121713
	1	85349	164832	92265	141334	94428	153761
skd-tree intersection search	0	69717	80573	69339	86039	70705	87551
	0.5	76593	90582	77039	96208	78658	98309
	1	80509	98744	81716	102319	83422	104464
skd-tree containment search	0	52914	61732	54848	67920	56358	69509
	0.5	53486	63072	55399	68931	56887	70531
	1	53773	66474	55641	69190	57092	70706

Storage Usage and Description				
Structures	$\theta_{object\ size}$	**Depth**	**Directory Pages**	**Data Pages**
Packed R-tree	0	3	92	1001
	0.5	3	92	1001
	1	3	92	1001
R-tree	0	4	184-192	1394-1423
	0.5	4	186-196	1399-1402
	1	4	182-193	1399-1402
Static skd-tree	0	3(11-12)	143-144	1000-1009
	0.5	3(11-12)	143-144	1000-1007
	1	3(11-12)	143-144	1000-1007
Dynamic skd-tree	0	4(14)	197-203	1407-1417
	0.5	4(13-14)	195-208	1416-1423
	1	4(14)	195-205	1417-1427

Table 4.7 Empirical results for page size = 128 bytes

Page Accesses for 10000 Queries							
Structures	$\theta_{object\ size}$	$\theta_{location}=0$		$\theta_{location}=0.5$		$\theta_{location}=1$	
		static	**dynamic**	**static**	**dynamic**	**static**	**dynamic**
R-tree	0	118396	222871	125028	209066	127909	212542
	0.5	123458	218393	139764	202825	142671	222084
	1	134429	219647	150280	220544	148935	214426
skd-tree	0	113482	127509	112992	128081	114818	131895
intersection	0.5	128088	146455	128421	148340	130900	151744
search	1	136535	157540	138261	160819	140899	165467
skd-tree	0	82852	92624	86147	95992	89515	99447
containment	0.5	84141	94709	87612	98065	90866	101667
search	1	84742	95642	88278	99036	91484	103405
Storage Usage and Description							
Structures	$\theta_{object\ size}$	**Depth**		**Directory Pages**		**Data Pages**	
Packed	0	5		403		2001	
	0.5	5		403		2001	
R-tree	1	5		403		2001	
	0	6		806-819		2753	
R-tree	0.5	6		811-822		2763-2772	
	1	6		806-835		2733-2794	
Static	0	5(13)		547-549		2000-2010	
	0.5	5(13)		547-549		2000	
skd-tree	1	5(13)		547-549		2000	
Dynamic	0	5-6(18-19)		771-785		2735-2792	
	0.5	5-6(17-19)		767-787		2767-2795	
skd-tree	1	5-6(17-20)		771-786		2762-2779	

While the location distribution becomes uniform, the performance of the dynamic R-tree [Gut84] improves relatively to the the packed R-tree [RoL84]. In fact, in some instances (Tables 4.3 and 4.4), the packed R-tree is more expensive than the dynamic R-tree. When the objects are more uniformly distributed, they are less overlapping. Thus the overlap of the covering rectangles in the internal nodes of the R-tree is also reduced. Generally, the packed R-tree performs better than the R-tree and requires a smaller directory and less data pages. Note that a reduction in directory and the number of data pages does not necessarily guarantee a more efficient indexing structure.

The performance of skd-trees, on the other hand, was more consistent and hence more predictable. As the skewness decreases, except for the largest page size (page size = 2048 bytes), the search cost increases slowly.

From the experiments, the following observations are made.

(1) When a page in an skd-tree is allowed to store more than one subtree, the storage requirement of the skd-tree becomes comparable to that of the R-tree. However, packed R-trees have smaller directories than those of static skd-trees.

(2) The query costs of the skd-tree are significantly smaller, especially when the page size is small relative to the number of nodes (i.e. the depths of both the skd-tree and R-tree are large).

(3) For the skd-tree, the *containment* search is far more efficient than the intersection search. As mentioned in Chapter 3, many operations (e.g. *contain* and *within*) can take advantage of such a search.

(4) The query cost of the skd-tree increases slightly as the skewness of locations and size of rectangles is reduced (the distribution becomes more uniform). The performance of the R-tree is more unpredictable. The containment search costs of the skd-tree is least affected by the nature of the data.

Due to the overlapping bounding rectangle used by the skd-tree, the skd-tree is not greatly affected by the location skewness. On the other hand, the sizes of the rectangles does affect the size of the bounding rectangles. Thus the dynamic skd-tree is affected more by the size skewness than the location skewness.

4.4 Further Comparisons

In order to make comparisons with other structures based on the kd-tree, we have implemented the 4d-tree and the mkd-tree. In some cases it was not possible to implement the mkd-tree with small page sizes as the density of some locations was be more than the page capacity.

4.4.1 Object Mapping Structures

The 4d-tree [BaK86] is a kd-tree [Ben75] in a 4-dimensional space. Regions in a 2-dimensional space are mapped into points in a 4-dimensional space and each region is represented as (x_1, x_2, y_1, y_2).

The region search presented in [Ben75] can be used for a spatial intersection search. Let the query region be (qx_1, qx_2, qy_1, qy_2). Then at each internal node one of the conditions, $x_1 \leq qx_2$, $x_2 \geq qx_1$, $y_1 \leq qy_2$, $y_2 \geq qy_1$, has to be used depending on the discriminator stored in that node to determine whether both subtrees or only one of the subtrees will need to be searched. In what follows, we present a search algorithm for the 4d-tree that is similar to that of the skd-tree in that a space is associated with each node. Although the 4d-tree is a 4-dimensional structure, the associated space is a 2-dimensional space. The algorithm is outlined as follows.

Algorithm 4d-Tree's Intersection Search

Input: node - a 4d-tree node; initially the root node.

                        br - current bounding space.

Output: a set of objects that intersect the query region *query_region*.

Comment: lo_br, hi_br - bounding rectangles to describe the LO and HI subspaces.

4D_SEARCH(node.hison_ptr, hi_br)

if node is a leaf **then**

    CHECK_LEAF(node);

    return;

lo_br = br; hi_br = br;

if node.disc = X_1 or node.disc = Y_1 **then**

    hi_br[node.disc] = node.disc_value;

else if node.disc = X_2 or node.disc = Y_2 **then**

    lo.br[node.disc] = node.disc_value;

if INTERSECT(lo_br, query_region) **then**

    4D_SEARCH(node.loson_ptr, lo_br);

if INTERSECT(hi_br, query_region) **then**

    4D_SEARCH(node.hison_ptr, hi_br);

end 4D_SEARCH.

The important part in the above algorithm is the determination of the subspaces that bound the objects in the LO and HI subtrees. Traversal starts at the root with the map as the associated space. The LO subtree contains objects whose X_1 coordinate is less than the *disc_value*, and the HI subtree contains objects whose X_1 coordinate is greater than the *disc_value*. The X_1 values of the HI subspace are bounded below by the *disc_value* and this fact can be used to reduce the subspace associated with the HI subspace. However, it is not possible to reduce the size of the LO subspace. Suppose the original map space is (x_1, x_2, y_1, y_2). Then the LO subspace is the same as that of the root node while the HI subspace is $(disc\_value, x_2, y_1, y_2)$. At the next level, the HI subspace remains unchanged, but for the LO subspace X_2 is bounded by the current discriminator value. When processing a query, it is quite common that both subtrees of a node will be searched. Figure 4.2 illustrates the case where both subspaces have to be searched.

For the experiments, we use the skd-tree paging strategy to page the 4d-tree directory. The experiments were then conducted with the same data and query sets used in the previous section.

The results are presented in Table 4.8.

As anticipated in Chapter 2, the intersection search is not efficient because of the inability to prune the search space. Less information is stored in a node compared to the skd-tree resulting in a smaller directory size. The results show that the performance of the 4d-tree is not competitve with that of the skd-tree.

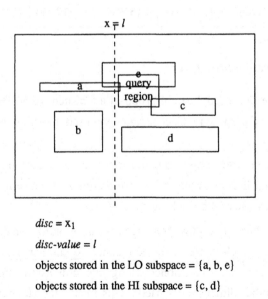

$disc = x_1$

$disc\text{-}value = l$

objects stored in the LO subspace = {a, b, e}

objects stored in the HI subspace = {c, d}

Figure 4.2 A 4d-tree objects distribution

Table 4.8 Empirical results for 4d-trees

Page Accesses for 10000 Queries							
Page size	$\theta_{object\ size}$	$\theta_{location}=0$	$\theta_{location}=0.5$	$\theta_{location}=1$	Depth	Dir. Pages	Data Pages
2048	0	149338	165784	169661	1-2(8)	1-2	167-172
	0.5	150289	160071	177011	1-2(8)	1-2	166-175
	1	166397	162333	165028	1-2(8)	1-2	166-170
1024	0	236555	248720	249634	2(10)	5-6	340-345
	0.5	244222	255218	259177	2(10)	5-6	338-350
	1	249037	262019	264257	2(9-10)	5-6	333-350
512	0	413454	359474	385645	2(12)	21-24	681-704
	0.5	417710	373612	386020	2(12-13)	21-24	679-684
	1	435745	385036	406042	2(11-13)	21-23	677-693
256	0	630260	623095	646732	3(15-16)	84-85	1406-1416
	0.5	682130	644469	679616	3(15-16)	83-84	1406-1421
	1	710384	668988	706992	3(14-16)	81-86	1406-1430
128	0	939013	1067814	1080776	4-5(17-20)	313-318	2773-2788
	0.5	972753	1123900	1121630	4-5(17-19)	315	2746-2791
	1	1014163	1156310	1155997	4(17-19)	311-319	2760-2781

4.4.2 Object Duplication Structures

We now consider the object duplication proposal advocated in [MHN84]. In [MHN84], no paging strategy is proposed. Once again, we used the paging strategy that was proposed for the skd-tree.

Highly dense data poses a big problem for structures employing object duplication. Example in Figure 4.3 illustrates this point. No matter where a split is introduced, there will be 4 rectangles in one of the resulting areas.

To overcome this problem, overflow data pages are introduced. That is, a leaf node may point to a data page, which in turn may be chained to another data page and so forth. Further, if a space cannot be partitioned along one dimension into two subspaces, a partition along the other dimension is tried. This means the strategy for partitioning a subspace along the dimension where the subspace has a longer interval advocated in

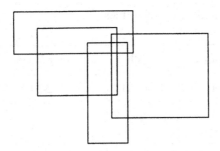

Figure 4.3 Highly dense data

[MHN84] is sometimes not applied. Tables 4.9a and 4.9b show the results of the mkd-tree with overflow data pages for the data and the queries used previously.

Table 4.9a Query retrieval for mkd-trees

Page Accesses for 10000 Queries							
Page size	$\theta_{object\ size}$	$\theta_{location}=0$		$\theta_{location}=0.5$		$\theta_{location}=1$	
		page access	objects duplicated	page access	objects duplicated	page access	objects duplicated
2048	0	45180	155769	29945	530	29840	472
	0.5	**	**	30038	1187	30066	1131
	1	**	**	32347	1806	30189	1789
1024	0	67811	128122	36458	783	36361	670
	0.5	**	**	37188	1864	30038	1187
	1	**	**	38402	3827	37316	3063
512	0	70928	75111	52419	1253	52677	970
	0.5	76353	74971	54761	4689	54793	2823
	1	**	**	58456	14690	54916	5796
256	0	89408	22291	63896	2599	65043	1594
	0.5	97207	19291	84321	10877	81912	5889
	1	**	**	99886	26563	89725	18231
128	0	116487	4810	109684	3966	111889	2878
	0.5	132134	3608	133951	9644	139555	10611
	1	**	**	160833	12871	155400	17581

Table 4.9b Storage requirement for mkd-trees

Page size	$\theta_{\text{object size}}$	Depth	Directory Pages	Data Pages
2048	0	2-3(9-91)	2-29	185-3261
	0.5	2(9-10)	3	209-210
	1	2(10)	3	229-231
1024	0	2-6(11-62)	10-81	394-5817
	0.5	2(12-15)	12-13	478-489
	1	2(12-19)	14-15	551-593
512	0	3(14-39)	44-211	857-7728
	0.5	3(14-38)	57-216	1109-8462
	1	3(15-30)	71-88	1422-2048
256	0	3-4(17-29)	200-418	1990-6979
	0.5	4(18-29)	304-367	3024-6923
	1	4(22-29)	503-558	5152-6702
128	0	5(22-24)	746-917	4679-6730
	0.5	5(22-24)	567-1368	6139-8505
	1	5(23-25)	1319-1667	11166-12731

The results show that the mkd-tree with overflow data pages supports efficient query retrieval for spatial objects that are relatively small and uniformly distributed. However, the trees become highly unbalanced for skewed data. Objects can sometimes be duplicated up to 15 times. The amount of duplication is determined by the distributions of object sizes and page sizes. Due to the large amount of object duplication, we did not run the simulation program for some highly skewed data (** in table).

To summarize, the object duplication method produces an efficient structure if the spatial objects are relatively small and evenly distributed. However, the major problems associated with this method is that the duplication of objects can be severe for skewed distributions and for large objects.

We attempted to implement the R^+-tree. The difficulties we encountered concern the problem of dead space (see Chapter 2 Section 2.5.11) and the difficulty in determining a suitable split. As discussed in Section 2.5.11, splits may propagate and the resulting R^+-tree can be very sparse.

Another key issue in partitioning a data page of the mkd-tree (or a leaf node of the R^+-tree) is that not only the number of objects being duplicated (or partitioned) must be minimized but the difference in the numbers of objects in the two resulting pages must also be minimized. No simple algorithm can achieve both these goals simultaneously.

4.5 Summary

The comparison between the skd-tree and the R-tree showed that the skd-tree is an effective data structure for supporting both containment and intersection searches. For the implementation of the skd-tree suggested in Section 4.4.3, the dynamic skd-tree has similar storage overhead, more efficient intersection search and for more effective containment search. For the static implementation of the skd-trees and the R-trees, the containment search of the skd-tree was more efficient and the other performance measurements were similar. Other extensions of the kd-tree proposed in the literature based on object duplication and object mapping appear to be less effective than the skd-tree.

Chapter 5

Query Optimization

5.1 Motivation

The query optimizer is an important component of a relational database management system. Its main task is to produce an efficient query evaluation plan for a given query. However, most optimizers in existing systems have been designed for conventional data models which are not capable of supporting applications like CAD/CAM or geographic data analysis. This has led to the current trend of designing extensible DBMSs [BaM86, Fre87b, GrD87, LMP87, Loh87]. Although, it has been shown that this approach works for simple cases [GrD87] it remains to be proven that the optimizers produced by these extensible DBMS provide efficient query evaluation for complex applications such as geographic information processing. In this chapter, we investigate an alternate method in which an existing DBMS is supplemented by new indexing structures and evaluation subsystems in order to process spatial queries. An objective of the approach is to take advantage of existing techniques for aspatial query evaluation while providing, at low costs, efficient implementation of spatial indexing techniques for general queries.

It is highly desirable that a GIS provides a storage and information processing architecture that integrates both the aspatial (attributes) and spatial components of the database. In such an integrated system, a user would be presented with a single query language, capable of expressing selection criteria that include both spatial and aspatial qualifications. Further, the system would support efficient execution of these hybrid queries. An example of such a query language is GEOQL which augments SQL with spatial operators to provide an interface that supports both spatial and aspatial query predicates.

Since conventional database management systems provide efficient storage and retrieval of attribute-based data, one approach to the design of a GIS is to extend an existing database system to include spatial and aspatial operators. Alternatively, a special purpose GIS, which requires full implementation, could be constructed to provide the full range of facilities. We opted for the first approach as it requires less effort in implementation and hence, less times and resources.

Consider the following query.

Example 5.1

Find all "Wellington" roads that intersect railway.

```
SELECT    railway.name
FROM      road, railway
WHERE     road.name = 'Wellington' and
          road intersects railway.
```

Queries such as this are very common in a GIS, and the response to such queries depends on the ability of the optimizer to find an efficient query evaluation plan. Without considering how selections and projections may be performed, there are several global strategies for executing the above query. After retrieving all roads with the name 'Wellington' using conventional selection techniques, two techniques exist for finding intersecting railways. A spatial index can be used to reduce the number of railway entities that must be processed, or else each railway must be examined in turn. If the *road* relation is very large and no index exists in the attribute *name*, it may be more efficient to perform the intersection test first using a spatial index. Ultimately, an optimizer must be able to make a choice among these schemes and select the low level operations.

A proposal of a general method for extending an existing DBMS so that queries involving spatial operators can be executed efficiently is given in this chapter.

5.2 Extending Techniques

5.2.1 Overview of An Extended System Architecture

Operations such as intersection and adjacency require the examination of spatial entities which may be irregularly shaped. One of the techniques [OMS87] to reduce the expensive testing that is involved is to associate a minimum bounding rectangle (MBR) with each entity. For instance, two objects do not intersect if their MBRs do not intersect. To evaluate a general spatial condition on two MBRs, four coordinate comparisons are necessary. To prune the search space, only **spatially close** objects should be examined, and indexing structures discussed in previous chapters are designed for such purposes. As a conventional DBMS does not directly support either of these structures or the necessary spatial operations, a *spatial processor*, is required to evaluate the spatial predicates both with and without the use of spatial indexes. This subsystem interfaces with the SQL backend through which it accesses the database, and performs necessary aspatial operations. The architecture of the extended system is illustrated in Figure 5.1. The *graphical display* unit which accesses graphical data provides the display of retrieved data in the form of maps.

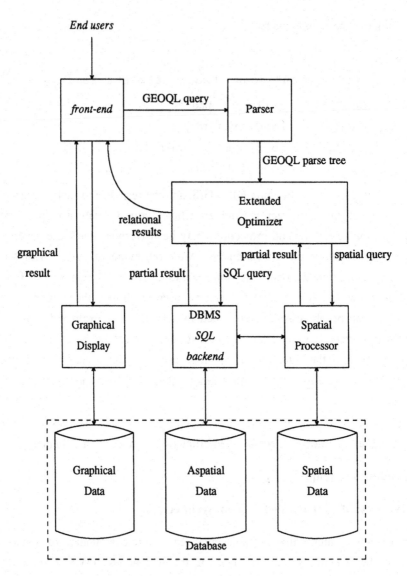

Figure 5.1 A system architecture for a GIS

5.2.2 Extended Optimization

Making use of an existing SQL backend is our primary intention. To this end, a GEOQL query is broken into subqueries that are either totally spatial or aspatial so that totally aspatial subqueries can be executed by an existing SQL backend, and the spatial subqueries can be executed by the spatial processor. Once the subqueries are formed, they are executed in an order that minimizes the overall query cost. The strategy consists of the following four major steps:

(1) *Logical Transformation*: As part of the optimization, the query tree produced by parsing the initial query is rearranged so that the new representation is more amenable to efficient evaluation. As well as including conventional logical transformations (e.g. redundancy removal), a GEOQL query must be restructured so that the use of the spatial indexing is made possible during optimization.

(2) *Decomposition*: The parse tree produced by stage 1 is partitioned into subtrees which are either totally spatial or aspatial. Each subtree represents a subquery which must be executed by the query processor. The aspatial subqueries will be executed by the existing SQL backend and the spatial subqueries which cannot be processed by the SQL backend can be executed by the spatial processor.

(3) *Plans Formulation and Selection*: From the set of subqueries obtained in the previous step, different orders are considered. The best or cheapest is chosen among all subquery sequences.

(4) *Plan Execution*: From the chosen subquery sequence, SQL subqueries are passed to the SQL backend and spatial subqueries are passed to the spatial processor.

The strategy will be described in detail in Section 5.4. Figure 5.2 shows the extended optimization model.

5.3 Classification of GEOQL Queries

Queries can be categorized into several classes according to the query complexity [BeC81, HeY79], predicate types and nesting involved [Kim82]. Based on the degree of nesting and the presence of spatial predicates, we partition GEOQL queries into four classes: unnested aspatial (UA), unnested spatial (US), unnested spatial and aspatial (USA), and nested queries (NSA).

A UA query has no spatial predicates and hence is a conventional SQL query. As the extended subsystem is assumed to exist on top of an existing DBMS that supports SQL, UA queries do not require preprocessing and are processed by the SQL backend directly.

Unnested queries which contain only spatial predicates in the WHERE clause are termed US queries. These queries consist of spatial predicates of the form R_i *geo_op* R_j, where R_i and R_j are geographic entity classes with an associated base relation or temporary relation. The associated relation will have as the primary key the *gid* attribute, values of which uniquely identify a spatial object. Each tuple of this relation describes the aspatial characteristics of an object, and the *gid* is also used to retrieve the full spatial description of the object from data object files as defined in Chapter 3. US queries are executed by the spatial processor.

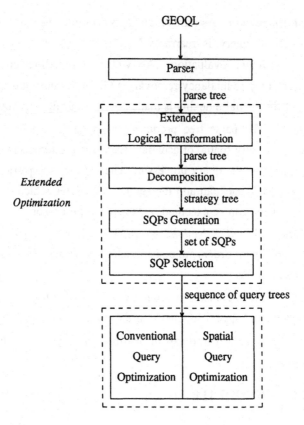

Figure 5.2 An extended optimization strategy

Associated with each entity class is a *spatial data structure* whose role is to facilitate retrieval based upon proximity. Data structures described in Chapters 2 and 3 have been proposed for such purposes. Although the skd-tree has been proposed and shown to be an efficient indexing structure, for the purpose of generality, we shall not preclude the use of other structures.

A US query may have more than one spatial predicate in its qualification. To process such a query, the query is decomposed into a sequence of US queries, each of which only contain a single spatial predicate. The spatial structure is used to facilitate the evaluation if one of the relations involved is a base relation; otherwise pairwise tuple evaluation is necessary. The partial results returned by the spatial processor are in the form of relations which may contain multiple *gid* attributes each describing an individual object. For subsequent spatial evaluation on the temporary relation, a predicate of the form R_1 *geo_op* R_2 must be rewritten into an internal representation of the form *R.gid1 geo_op R.gid2* to avoid confusion in these cases.

Typically, queries will not only select data based on the spatial relationships but also specify some aspatial characteristics. We classify this class of queries as USA, which contain a non-nested combination of spatial and aspatial predicates. A USA query can be decomposed into a sequence of UA and US subqueries. Additional UA queries are introduced to combine the intermediate results (see Section 5.8).

SQL is a block-structured language that allows nested queries. GEOQL is a consistent extension of SQL that preserves the basic syntactic structure and allows spatial predicates to occur in the WHERE clause of an outer or inner SELECT expression. Therefore GEOQL queries may be nested.

The ability to express complex selection criteria by *nesting* of query blocks is one of the most powerful features of SQL. However, as mentioned in Chapter 2, conventional techniques used in evaluating a nested query (e.g. as in System R) can be very inefficient [GaW87, Kim82]. Query transformation algorithms which transform a nested query into a logically equivalent sequence of unnested queries to improve the efficiency have been proposed [GaW87, Kim82]. As GEOQL spatial predicates do not allow subqueries as operands, the GEOQL language introduces no new nesting complexity, and so the existing SQL unnesting algorithms may be applied with only minor modifications. The result of this unnesting is a sequence of USA, US, or UA queries. Unnesting is considered a preprocessing stage, and the optimization strategies that follow consider each unnested query in isolation; consequently the only scope for optimization between the nested components occurs in the unnesting phase.

5.4 An Introduction to the Extended Optimization

In this subsection, we give an intuitive description of the fundamentals behind the decomposition and partial result merging strategies proposed later.

Before we proceed any further, we define some of the terms used in what follows. In Chapter 2, we used $P_k(S_j)$ to denote kth predicate defined over a set of relations (jth set) S_j. In order to distinguish a spatial predicate from an aspatial predicate, we put a superscript s on a predicate (e.g. P_k^s) when the predicate is spatial and a when it is aspatial (e.g. P_k^a). When the type of predicate is not important to the context of discussion, we simply use P_k.

We use $T(P_k(S_k))$ to denote a table that satisfies the predicate $P_k(S_k)$ and $I(S_k)$ an identity function which denotes the extensional form of S_k. $ATT(S)$ is used to denote the set of attributes of relations defined in S. For a logical expression we write $log\_op(P_1(S_1), P_2(S_2), ..., P_n(S_n))$ to denote the disjunction or conjunction of the predicates $P_1(S_1), P_2(S_2), ...$, and $P_n(S_n)$, where log_op is *or* or *and* respectively.

5.4.1 Grouping of Predicates

Whenever permissible, several predicates involving the same relations are grouped together and considered as one entity (e.g. two predicates ($R_1.A$ = "constant1" and $R_1.B$ = "constant2") should not be evaluated separately). The inherent advantage in grouping the predicates involving the same relation(s) is that potential redundant page reads are effectively reduced if not totally avoided. Indeed, the grouping of predicates whose operands are not indexed with those for which indexes exist may lead to efficient evaluation strategies. Further, the grouping of predicates avoids unnecessary creation of subqueries by the decomposition strategy that will be introduced later.

Following the above discussion, an SQL predicate may refer to a conjunction or disjunction of simpler predicates involving the same relation(s). For example, a conjunction ($P_1^a(\{R_1, R_2\})$ *and* $P_2^a(\{R_1, R_2\})$) is treated as $P_3^a(\{R_1, R_2\})$, where P_3 is a symbolic name given to the compound predicate. Consider another example: ($P_1^a(\{R_1, R_2\})$ *and* $P_2^a(\{R_1\})$). These predicates are not necessarily grouped together since P_2^a may be used to restrict other spatial predicates.

Spatial predicates are left as simple predicates since the spatial processor has to examine them one at a time.

5.4.2 Answer Construction

In this subsection, we look at how an answer can be formed if predicates of a conjunction/disjunction are evaluated separately. The strategy used is the basis for the rewriting rules we introduce in Section 5.8.1.

During query evaluation a GEOQL query is transformed into an evaluation tree consisting of simple predicates connected by *boolean* operators. As in the relational calculus, the logical operators supported include *not, and* and *or*. Using standard transformations of boolean expressions, not operators are either removed from the evaluation tree or pushed as far down the tree as possible. The conjunctive (*and*) operator in relational calculus corresponds to the *cross* and *join* operators of relational algebra. The partial results of two conjuncted predicates must be merged using either an equi-join or a cross product operation depending on the referenced relations. Similarly, the disjunctive (*or*) operator in relational calculus corresponds to the *cross* and *union* operators of the relational algebra.

5.4.2.1 The Not Operator

The complement of an expression is always transformed into a logically equivalent expression for which the not operator is either removed totally (e.g. $not(R_1.A > R_2.B) \Rightarrow (R_1.A \leq R_2.B)$) or pushed as far down the evaluation as possible (e.g. $not(and(P_1, P_2)) \Rightarrow (or(not(P_1), not(P_2))))$.

5.4.2.2 The And Operator

Given a conjunction of predicates, the optimizer chooses a predicate to evaluate, then the next (possibly without using the previous result), and so on. If the optimizer chooses not to use the partial results of previously evaluated predicates in further evaluations, then the optimizer has to merge all partial results to obtain the answer to a query.

Consider the following expression:

$$R_1.A > constant1 \text{ and}$$
$$R_1.B = R_2.B \text{ and}$$
$$R_3.C = constant2.$$

To evaluate this expression, the result of the first predicate can be equi-joined with the result of the second predicate, and the result of the equi-join can then be crossed with the result of the third predicate. This sequence of steps may not occur in actual evaluation; our intention here is to show how should we merge the partial results for a correct answer if the predicates are executed separately.

So, given $P(S_i \cup S_j) = and(P_i(S_i), P_j(S_j))$, then

$$T(P) \equiv \begin{cases} T(P_i(S_i)) \bowtie T(P_j(S_j)) & \text{if } I \neq \emptyset \\ T(P_i(S_i)) \times T(P_j(S_j)) & \text{otherwise} \end{cases}$$

where $I = ATT(S_i \cap S_j)$.

It is assumed that I is the set of attributes common to both predicates and the join is an equi-join on the common attributes. Furthermore, if a base relation is referenced in both predicates as different instances (correlations), then both instances are treated as two different relations although both instances may be obtained with one read. This approach only involves attribute renaming.

5.4.2.3 The Or Operator

The result of a disjunction of two predicates referencing the same relations is the union of both partial results. If two predicates do not reference the same relations, then before a union is formed the partial result of each predicate must be crossed with those relations that are not involved in the predicate but are involved in the other predicate.

Consider the following expression:

$$R_1.A > \text{constant1 or}$$
$$R_1.B = R_2.B \text{ or}$$
$$R_3.C = \text{constant2.}$$

The partial result of the first predicate must be crossed with the relation R_2 before a union with the partial result of the second predicate can be performed. Then the result of the union must be crossed be R_3, and the partial result of the third predicate must be crossed with R_1 and R_2, so that a union of both results can be performed to yield the answer. In our notation, the operations are in the form: $\{[T(R_1.A > \text{constant1}) \times T(R_2)] \cup [T(R_1.B = R_2.B) \times I(R_3)]\} \cup \{I(R_1) \times I(R_2) \times T(R_3.C = \text{constant2}\}$. The above process is the same as firstly crossing all partial results with the relations not involved in its predicate, and then performing a union of all results.

Now let $P = \text{or}(P_i(S_i), P_j(S_j))$, then the result without any projection is defined as follows:

$$T(P) \equiv \begin{cases} T(P_i(S_i)) \cup T(P_j(S_j)) & \text{if } S_i = S_j \\ (T(P_i(S_i)) \times T(I(S_j \text{ minus } S_i))) \cup (T(P_j(S_j)) \times T(I(S_i \text{ minus } S_j))) & \text{otherwise} \end{cases}$$

Here we presume that the columns are rearranged so that when the union is performed, all operands are union compatible [Dat86].

The reasoning behind the *and* and *or* gives rise to the rewriting rules we introduce in Section 5.4.5.

In the following sections we describe each stage of optimization process.

5.5 Optimization Strategy

5.5.1 Parsing

GEOQL is an extension of SQL designed so that an existing SQL parser can be extended without major modification for GEOQL queries queries. For subsequent query transformation, it is highly desirable that the structure used to represent a parsed GEOQL query allows a great degree of freedom for restructuring. The parse tree used is a *multiway* tree, for which an internal node may have more than two children. An example of a complete representation of a GEOQL query is illustrated in Figure 5.3.

For the discussion below, we need only consider the predicate tree, i.e. that part of the parse tree representing the WHERE clause of the GEOQL query when the other parts of the parse tree are immaterial to the context of discussion. Further we consider a predicate tree as consisting of internal nodes representing logical operators (and/or) and external nodes representing atomic predicates.

```
Query:          Find all roads that are adjacent to a lake
                whose  usage is either  'recreational'  or
                'irrigation'.
GEOQL Query:    SELECT   road.name
                FROM     lake, road
                WHERE    (lake.usage = 'recreational' or
                         lake.usage = 'irrigation') and
                         lake is adjacent to road.
```

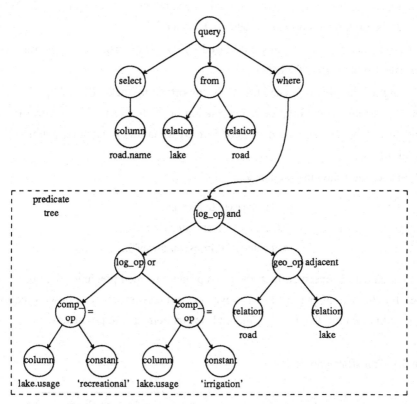

Figure 5.3 A GEOQL parse tree

5.5.2 Logical Transformation

Nested queries have been shown to be costly for query evaluation [GaW87, Kim82].
Thus nested queries are transformed into a partially order set of simpler unnested queries.
As GEOQL spatial predicates do not allow subqueries as operands, the GEOQL language
introduces no new nesting complexity, and so the existing SQL unnesting algorithms
[GaW87, Kim82] may be applied with only minor modifications. As a result of this

unnesting, the optimization strategies that follow consider each unnested query in isolation.

As in SQL, several GEOQL queries with different syntactic forms but with the same semantics may be formed. These alternative syntactic forms may exhibit different potential for query optimization, and greatly increase the complexity of optimization procedures. One of the preprocesses involved in query analysis is to convert all queries into some canonical form [Hal74, Hal76], with the objective of reducing the syntactic variants amongst semantically equivalent queries and yielding syntactic forms that are most amenable to the optimization heuristics. Assuming that a syntactically correct query is translated into a parse tree, the predicate subtrees corresponding to the WHERE clause may be initially simplified as stated in Chapter 2.

The precedence of predicates introduced by means of brackets can be removed by the two transformations in Table 5.1.

As argued in previous chapters, it is important to make use of spatial indexes whenever possible. Therefore, the representation of predicates in the predicate tree must not cause an early Cartesian product to be formed when this product may inhibit the use of spatial indexes.

Consider the following predicate.

> road is adjacent to lake and
> (lake.usage = "recreational" or
> road.name = "Princess Highway").

If the *or* predicate is executed first, the cross product formed precludes the use of spatial indexes for the spatial predicate. However the expression in the following equivalent form would not inhibit the use of spatial indexes on both spatial predicates.

Table 5.1 Transformation rules

Expression	Equivalent Form
Associative Rules	
$or(P_1(S_1), or(P_2(S_2), P_3(S_3)))$	$or(P_1(S_1), P_2(S_2), P_3(S_3))$
$or(or(P_1(S_1), P_2(S_2)), P_3(S_3))$	
$and(P_1(S_1), and(P_2(S_2), P_3(S_3)))$	$and(P_1(S_1), P_2(S_2), P_3(S_3))$
$and(and(P_1(S_1), P_2(S_2)), P_3(S_3))$	

> (road is adjacent to lake and
> road.name = "Princess Highway") or
> (road is adjacent to lake and
> lake.usage = "recreational").

Note also in this example that the restrictions may be used to reduce the cost of evaluating the spatial expressions and no cross product is involved.

The second objective of the transformation to reduce the number of terms in an expression by identifying identical terms, which indeed is part of a conventional transformation. Consider the following DNF expression.

> (lake.name ≠ "Eildon" and
> lake is adjacent to railway) or
> (lake.usage = "recreational" and
> lake is adjacent to railway).

The identification of common subexpressions would reduce the above expression to the following more efficient formulation.

> lake is adjacent to railway and
> (lake.name ≠ "Eildon" or
> lake.usage = "recreational").

As noted before, a conjunction or disjunction of atomic (simplest) predicates involving the same relation/s is always treated as a single predicate. Thus for the above expression, predicates of the form (P^s({lake, railway}) and P^a({lake})) result. The advantage of this representation is that the decomposition strategy (next section) would only have to consider two terms and create two subqueries, one for each predicate. In addition, the predicates expressed in such form would also avoid an extra merging of partial results and possibly enable the SQL backend to optimize the compound predicate (e.g. lake.name ≠ "Eildon" or lake.usage = "recreational") as a whole.

Notice that the two objectives, namely making use of spatial indexes and the identification of common subexpressions, are in conflict. Allowing the use of spatial indexes is given higher priority than reducing the number of terms. Thus, the transformation required on top of the existing conventional logical transformations can be stated as follows:

$$\text{and}(P_1^s(S_1), \text{or}(P_2(S_2), P_3(S_3))) \Rightarrow \text{or}(\text{and}(P_1^s(S_1), P_2(S_2)), \text{and}(P_1^s(S_1), P_3(S_3)))$$
$$\text{if } (S_1 \cap S_2 \text{ or } S_1 \cap S_3) = \text{true}$$

The above transformation is necessary, irrespective of the predicate types of P_2 and P_3. Note that the above transformation has no effect on the following expression.

lake.usage = "recreational" and

road.name = "Princess Highway" and

(lake is adjacent to road or

lake is adjacent to railway)

For this expression, the spatial predicates are not impeded from using spatial indexes and can be more efficiently evaluated by using the partial results of the restrictions. No transformation is necessary.

5.5.3 Decomposition

A well known technique used in query optimization is to decompose a query into simpler components to reduce the complexity of optimization. In general, the simpler subqueries are designed to reference fewer relations than the original query [WoY76, Yao79]. In our context, a USA (unnested GEOQL) query is decomposed into simpler US and UA subqueries to enable global ordering of subqueries such that an overall evaluation cost is minimized. However, there remains considerable choice in the extent to which the SQL predicates should be decomposed. Some of the criteria used to decide the extent to which an SQL subquery should be decomposed are listed below.

In order to reduce page accesses, the optimizer never breaks a query into multiple subqueries that might result in unnecessary rereading of the same table. Also, the subqueries that are produced should not require expensive operations, such as the formation of cross products, when the partial results are combined.

Consider the examples in Figure 5.4. In Figure 5.4a, a temporary relation created by P_i^a may affect the strategy for executing P_j^s. On the other hand, the partial result of an execution of P_3^a or P_4^a in Figure 5.4b can not affect the execution of P_5^s no matter what attributes are included in S_3 and S_4. Hence, for the predicate tree in Figure 5.4b, P_5^s, P_6 and or(P_3, P_4) would be detached as subqueries. P_1^a and P_2^a would form another two subqueries if either S_1 or S_2 intersected S_5; otherwise the conjunction of the two predicates would form just one subquery. A more complicated decomposition policy would form P_1^a and P_2^a as two subqueries if S_1 or S_2 intersected any of S_3, S_4, S_5 and S_6. However, this more complicated strategy could lead to severe increase in the cost of determining the optimal subquery sequencing. The strategy we have adopted is to form subqueries

(1) for each spatial predicate.

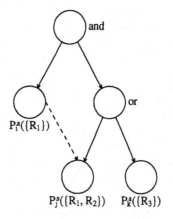

(a) An instance of downward propagation

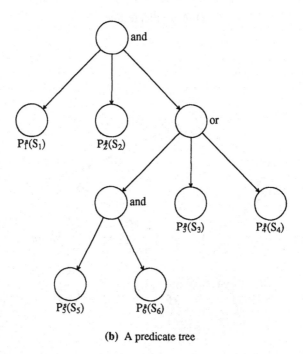

(b) A predicate tree

Figure 5.4 Grouping of predicates

(2) for each SQL predicate (which may be compound) that is related to a spatial predicate by a conjunction and references the relations involved in the spatial predicate.

(3) for those remaining SQL predicates in a logical expression which do not satisfy condition (2).

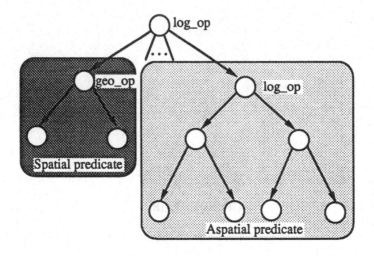

(a) A predicate tree

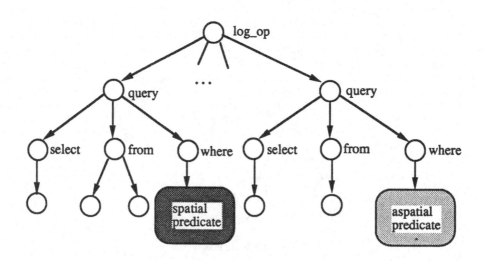

(b) A corresponding strategy tree

Figure 5.5 Mapping of a predicate tree to a strategy tree

The decomposition scheme maps a GEOQL expression $\log\_op(P_1, P_2, ..., P_k)$ into a query expression $\log\_op(E_1, E_2, ..., E_n)$, where E_i is either a query expression or an US or UA query $Q_k(S_k)$. More precisely, it is a mapping of a predicate tree into a tree whose internal nodes are logical operators, and external nodes are queries (pointers to corresponding query trees). We call the data structure formed by the decomposition algorithm the *strategy tree*. Figure 5.5 illustrates the mapping produced by the decomposition scheme. With such representation, the dependency of the subqueries is captured without any ordering.

A join, $R_1.A = R_2.A$, where one of the relations (R_2) does not participate in any other predicates and the target list (answer) is implicitly a restriction. Hence, we regard such a join as a restriction in our decomposition algorithm.

The abstract form of the decomposition algorithm is presented below. A more detailed form of the algorithm can be found in Appendix B.

Algorithm Decompose

Input: A predicate tree.
output: A strategy tree.

Traverse the predicate tree in depth first left to right order;
At each logical node
    Form a US (spatial) query for each spatial predicate;
    Form a UA (aspatial) query for a predicate subtree that does not contain any spatial predicate and the referenced relation(s) is(are) also referenced by a spatial predicate related by a conjunction.
    Form a UA query for those remaining aspatial predicate subtrees if the sibling subtrees have been used to form subqueries.
    Construct the corresponding subtree for the subqueries formed above in the strategy tree;

As an example of the decomposition algorithm, the following contrived query (disregard its semantics) is provided to illustrate the power of the decomposition strategy:

```
SELECT    road.name, railway.name
FROM      road, region, railway
WHERE     road.name ≠ region.name and
          road is adjacent to region and
          railway.name ≠ 'R10' and
          (region.name = 'Monash University' or
           region.name = 'RMIT').
```

Four subqueries formed by the decomposition strategy are given in Figure 5.6.

```
Q₁:   SELECT    *
      FROM      region, road
      WHERE     road is adjacent to region.


Q₂:   SELECT    *
      FROM      region
      WHERE     region.name = 'Monash University' or
                region.name = 'RMIT'.


Q₃:   SELECT    *
      FROM      road, region
      WHERE     road.name ≠ region.name.


Q₄:   SELECT    *
      FROM      railway
      WHERE     railway.name ≠ 'R10'.
```

Figure 5.6 Subqueries formed by the decomposition

The strategy tree for the above query is an internal node with an *and* logical operator pointing to four offspring subqueries.

Note that the Q_1, Q_2 and Q_3, and Q_4 are formed under conditions 1, 2 and 3 respectively.

It should be noted that the decomposition algorithm only decomposes a query into multiple subqueries, it does **not** specify any execution order for the subqueries. It is possible to arrange the subqueries in the strategy tree in any order so long as the final answer is correct. Hence the final forms of the SELECT clauses of subqueries cannot be defined until the full ordering is known. In Figure 5.6, the SELECT clauses of Q_1, Q_2, Q_3 and Q_4 have not been set to any particular attributes. These four subqueries are not independent; they are related by conjunctions and the strategy tree captures this dependency. Through this dependency, the partial results of subqueries may be merged or used by other subqueries.

5.5.4 Subquery Sequencing

5.5.4.1 Results Formation

Since a strategy tree is semantically equivalent to the predicate tree from which it originates, the subqueries in the strategy tree and their corresponding predicates in the original query tree exhibit the same dependencies. Thus, the principles of result merging mentioned in Section 5.4 can be adopted for ensuring the correctness of the result of subquery executions. While ordering the subqueries, the dependency captured by the strategy tree structure can be well expressed by the *introduction of extra SQL queries* and/or *terms rewriting*. The extra queries are used to merge multiple partial results (temporary relations), while the technique of terms rewriting propagates the partial result of a query to another query. Three result rewriting rules are introduced in this section to ensure the correctness of the final result. These rules are *mutually exclusive*, and the conditions for which the rules are applied are described in Table 5.2.

Table 5.2 Result rewriting rules

Expressions	Conditions	Rules
$and(Q_1(S_1), Q_2(S_2))$	$S_1 \cap S_2$	1
$and(Q_1(S_1), Q_2(S_2))$	$S_1 \pitchfork S_2$	2
$or(Q_1(S_1), Q_2(S_2))$	NIL	3

If two subqueries are connected by a conjunction, reference some common relations and are executed independently, then the partial results must be merged by a natural join. However, if a subquery (subquery Q_a) uses the partial result of the other subquery (subquery Q_b), and the latter subquery is executed first, then no natural join is required since the result that is produced by Q_a also satisfies the selections in Q_b.

Result Rewriting Rule 1. Given an expression $and(Q_1(S_1), Q_2(S_2))$ and the precondition $S_1 \cap S_2 \neq \emptyset$. Let $I = S_1 \cap S_2$ and let $C \subseteq I$. Suppose Q_1 is executed before Q_2, which results in a temporary relation T_1, then for each $R \in C$ ($C \neq \emptyset$), replace R by T_1 in Q_2, and

(1) if $C = I$, no further rewriting is necessary.

(2) if $C \subset I$ or $C = \emptyset$ and suppose T_2 is the temporary relation produced by Q_2 and A_1, ..., A_n are the common unqualified attributes in I, then the following SQL query (an equijoin) is introduced to join merge the partial result:

$$
\begin{array}{ll}
\text{SELECT}^\dagger & * \\
\text{FROM} & T_1, T_2 \\
\text{WHERE} & T_1.A_1 = T_2.A_1 \text{ and} \\
& T_1.A_2 = T_2.A_2 \text{ and} \\
& \quad ... \\
& T_1.A_n = T_2.A_n.
\end{array}
$$

Although the above rewriting rule is formulated with only two terms, it is however applicable for a general case by consecutively considering two terms at a time. It is worth mentioning that one tends to think for the second case that, when $C \subset I$, an equijoin of the attributes in I-C is sufficient. This nevertheless does not guarantee that attribute values of attributes in I-C and attribute values of attributes in C satisfy the predicates of both subqueries at the same time.

Using a restricted instance of a base relation or a base relation with indexes is determined by the optimizer to minimize the processing cost. That is, the choice of the number of relations in C is determined by the query optimizer. On the one hand, result rewriting can lead to smaller (restricted) relations being propagated to the other subqueries. On the other hand, the use of indexes in evaluating these subqueries is then not possible.

Proposition 1 establishes the equivalence of the original query and its subqueries sequence.. At this stage, all attributes are assumed to be selected in the SELECT clause, which are dealt with later. For illustration, the attributes of R_i are assumed to consist of $\{A_1, A_2\}$.

---

† since this is meant to be a natural join, only one of the joining attributes would be selected.

Q_1:	SELECT	*
	FROM	R_i, R_j
	WHERE	$P_1(R_i, R_j)$ and $P_2(R_i)$.
Q_2:	SELECT	*
	FROM	T_2, R_j
	WHERE	$P_1(T_2, R_j)$.
$T_2=$	SELECT	*
	FROM	R_i
	WHERE	$P_2(R_i)$.
Q_3:	SELECT	*
	FROM	T_{3a}, T_{3b}
	WHERE	$T_{3a}.A_1=T_{3b}.A_1$ and $T_{3a}.A_2=T_{3b}.A_2$.
$T_{3a}=$	SELECT	*
	FROM	R_i
	WHERE	$P_2(R_i)$.
$T_{3b}=$	SELECT	*
	FROM	R_i, R_j
	WHERE	$P_1(R_i,R_j)$.

Proposition 1. Result rewriting rule 1 does produce the correct answer, that is, Q_1, Q_2 and Q_3 are equivalent.

Proof. Q_1 can be thought of as first fetching a tuple of R_i. If P_2 is true, fetch all the tuples of R_j such that P_1 is true. Then the result is projected as the final answer. This is operationally the same as Q_2, except Q_2 will have the partial result of R_i for which P_2 is true stored as the temporary relation T_2. Then for each tuple of T_2 and for all tuples of R_j, the result is formed if P_1 is true. Another way which is similar to Q_2, is firstly evaluating P_1 and then, based on the partial results, P_2 is evaluated. Alternatively, P_1 and P_2 are evaluated independently, and a join of two temporary relations on the common attributes is performed. This is expressed as Q_3. Hence the three queries produce the same answer. □

In the next case, we can view the evaluation of a query as being performed by firstly forming the Cartesian product of all the relations that are referenced, and then evaluating the predicates.

When two conjuncted subqueries reference different relations, the order of their evaluation is not important to cost saving. However, the partial result must be propagated to become part of the final answer. Subqueries Q_4 in Figure 5.6 is a case in point. The partial result of Q_4 must participate in the final answer.

Result Rewriting Rule 2. We are given an expression and$(Q_1(S_1), Q_2(S_2))$, and a condition $S_1 \pitchfork S_2$. Suppose that Q_1 is evaluated before Q_2, then put the temporary relation T_1 produced by Q_1 in the FROM clause of Q_2.

To show that result rewriting rule 2 ensures the correct answer of any subqueries sequences, consider the following query and its equivalence.

Q_4:	SELECT	*
	FROM	R_i, R_j, R_k
	WHERE	$P_1(R_i, R_j)$ and $P_2(R_k)$.

Q_5:	SELECT	*
	FROM	R_i, R_j, T_5
	WHERE	$P_1(R_i, R_j)$.

$T_5=$	SELECT	*
	FROM	R_k
	WHERE	$P_2(R_k)$.

Proposition 2. Result rewriting rule 2 is correct, that is, Q_4 and Q_5 are semantically equivalent.

Proof. Both predicates P_1 and P_2 can be evaluated independently, and the answer is the cross product of the partial results of P_1 and P_2. This is the same as forming the cross product of R_i, R_j and R_k, and then retaining the tuples for which P_1 and P_2 evaluate to true at the same time. For Q_5, P_2 is evaluated first and the result is stored as T_5. Then for each tuple of R_i, and for each tuple of R_j, the two tuples are merged if P_1 is satisfied and the resultant relation is then crossed with T_5. As P_1 and P_2 can be evaluated independently, the sequence of evaluation is immaterial. $\square$

Notice that Q_5 has the similar effect as the following query.

Q_6:   SELECT   *

        FROM T_{6a}, T_{6b}.

$T_{6a}=$ SELECT   *

        FROM R_i, R_j

        WHERE $P_1(R_i, R_j)$.

$T_{6b}=$ SELECT   *

        FROM R_k

        WHERE $P_2(R_k)$.

Two disjuncted subqueries that reference different sets of relations must be made union compatible and hence, each partial result must be "crossed" with the relations that are not referenced by that subquery.

Result Rewriting Rule 3. Given an expression or$(Q_1(S_1), Q_2(S_2), ..., Q_k(S_k))$, we let $S = S_1 \cup S_2 \cup ... \cup S_k$. The FROM clause of Q_i ($i = 1 .. k$) is rewritten with the set S as the referenced relations. Suppose T_i is the temporary relation produced by Q_i, then the result is formed by:

$$T_1 \text{ UNION } T_2 \text{ UNION ... } T_k.$$

The following two queries are formed to show the correctness of the above result rewriting rule.

Q_7:   SELECT   *

        FROM R_i, R_j, R_k

        WHERE $P_1(R_i, R_j)$ or

                    $P_2(R_j, R_k)$.

Q_8: T_{8a} UNION T_{8b}.

$T_{8a}=$ SELECT   *

        FROM R_i, R_j, R_k

        WHERE $P_1(R_i, R_j)$.

$T_{8b}=$ SELECT   *

        FROM R_i, R_j, R_k

        WHERE $P_2(R_j, R_k)$

Proposition 3. Result rewriting rule 3 is correct, that is, Q_7 and Q_8 are equivalent.

Proof. Q_7 can be thought of as firstly forming the Cartesian product of all the relations and duplicating them as two temporary relations. Then evaluate P_1 on one of the temporary relations and P_2 on the other. The results obtained by both evaluations are then unioned to form the final answer. This is equivalent to Q_8. □

Notice that the first rewriting rule does not handle a general case like $log\_op(E_1, \ldots, E_n)$, where E_i may be a subquery or an expression. Consider the example: $and(Q_1(R_1)$, $or(Q_2(R_1, R_2), Q_3(R_1))$ and suppose Q_1 is the first to be evaluated. Suppose we choose to to use the partial result of Q_1 in evaluating Q_2 but not Q_3 for some reason (like using indexes). Then a natural join between the results of the *or* expression and the results of Q_1 is required. However, if both Q_2 and Q_3 use the partial result of Q_1, then the partial result can be considered as being propagated to the *or* expression and hence no extra natural join is required.

The partial result relation of one subquery may be used as a substitute for a base relation of another subquery if the first common ancestor of both subqueries is an *and* node. In Figure 5.7, any of the partial results of Q_1^a, Q_2^a, Q_3^a and Q_4^a may be used by Q_5^a and Q_6^a. A child of an *and* node may use its brothers' results. Suppose insteads of Q_6, we have an *and* node with two child subqueries. These two subqueries may then use partial result of each other and the partial results of Q_i ($1 \le i \le 4$) but not of Q_5^a.

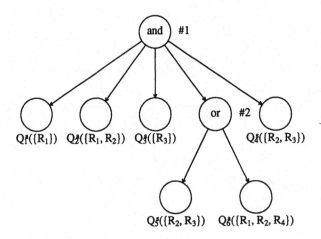

Figure 5.7 Use of partial results

5.5.4.2 SQP Formulation

For n subqueries, n! distinct sequences of subqueries - subquery plans (SQPs) can be formulated by the brute force method. Without pruning techniques, the optimizer would be unacceptably slow. Hence a rule-based and heuristic strategy is required to reduce the number of SQPs considered. Following the conventional DBMSs, a rule-based optimizer should employ certain "*rules*" or "*musts*", to avoid the generation of sequences that are obviously ineffective, and heuristics to further prune the search space. In this section, we describe how we generate and prune the search space for finding the subquery sequence that costs least.

5.5.4.2.1 Spatial Evaluations

Important to the cost of an SQP is the way in which a spatial subquery is evaluated. Suppose after decomposition, the subqueries $Q_h^g(\{R_m, R_n\})$, $Q_i^a(\{R_m\})$, $Q_j^a(\{R_n\})$, and $Q_k^a(\{R_m, R_n\})$ are a subset of the subqueries formed, and suppose they are related by conjunctions. The spatial subquery may be evaluated as one of the sequences outlined in Table 5.3[†].

Table 5.3 Spatial evaluation strategies

Subquery Sequences	Spatial Evaluations
... $Q_h^g(\{R_m, R_n\})$ ...	**for** each tuple in R_m (or R_n) **do** search the skd-tree of R_n (or R_m)
... $Q_i^a(\{R_m\})$ ... $Q_h^g(\{T(Q_i^a), R_n\})$ ...	**for** each tuple in $T(Q_i^a)$ **do** search the skd-tree of R_n
... $Q_j^a(\{R_n\})$ ... $Q_h^g(\{R_m, T(Q_j^a)\})$ ...	**for** each tuple in $T(Q_j^a)$ **do** search the skd-tree of R_m

... $Q_i^a(\{R_m\})$
 ... $Q_j^a(\{R_n\})$
 ... $Q_h^g(\{T(Q_i^a), T(Q_j^a)\})$
 ...
	for each tuple in $T(Q_i^a)$ **do**
	 for each tuple in $T(Q_j^a)$ **do**
	 do the spatial evaluation

... $Q_i^a(\{R_m\})$ ... $Q_j^a(\{R_n\})$ ... $Q_k^a(\{T(Q_i^a), T(Q_j^a)\})$ ... $Q_h^g(\{T(Q_k^a)\})$ ...	**for** each tuple in $T(Q_k^a)$ **do** do the spatial evaluation

---

† Here we assume that the spatial predicates are of those commutative types (i.e. geo_obj1 geo_op geo_obj2 $\Rightarrow$ geo_obj2 geo_op geo_obj2); otherwise, less alternatives are possible.

In Chapter 4, we have shown that the cost of using indexing structures is relatively much cheaper than scanning the entire database sequentially. Therefore, it is beneficial to use existing spatial indexing structures whenever possible. In the case where both entities are indexed, the smaller relation will be used as the outer relation and for each object of this relation, the indexing structure of the other relation is searched. For GEOQL spatial predicates, the use of spatial indexes may be restricted to a particular operand of a predicate (e.g. for *contain* and *within* operators).

5.5.4.2.2 Search Space Pruning

The set of subqueries formed by the decomposition strategy is generally small. However, it is important not to consider any alternatives that are obviously ineffective. Heuristic techniques are used to generate SQPs that are efficient. A relaxed heuristic uses less rules and hence produces more alternatives than those produced by a stricter heuristic which has more rules. A disadvantage of a stricter heuristic is that the likelihood of a good alternative being rejected is higher.

The following rules are proposed to reduce the subquery sequences being generated and thereby reducing the search space for finding the best sequence.

Heuristic Rule 1. For an expression $and(Q_1(S_1), ..., Q_i(S_i), ..., Q_n(S_n))$, if $S_i \pitchfork (S_1 \cup ... \cup S_{i-1} \cup S_{i+1} \cup ... S_n)$ then $Q_i(S_i)$ is executed just before the result is required.

Discussion. The execution of Q_i does not affect any intermediate results of Q_j for $1 \le j \le n$ $(i \ne j)$, therefore the sequencing of its execution is not important. To avoid rereading the partial result of Q_i which is not required till the formation of the final answer, the merging of the partial result can be postponed until n-2 sibling subqueries have been evaluated. One of n-1 queries, Q_j $(j \ne i)$, is not executed before Q_i so that the partial result of Q_i may be propagated to the final answer through that query (as in Result Rewriting 2). $\square$

The above rule is applicable for a general case in which there is more than one independent subquery. However, for each *and* node of the strategy tree, the decomposition strategy will produce at most one query for which the referenced relations do not exist in the other conjuncted queries.

Heuristic Rule 2. For an expression $or(Q_1(S_1), ..., Q_n(S_n))$, the order of execution of subqueries is immaterial to cost saving. Hence the queries may be executed in any specific order.

Discussion. Since the subqueries are independent, that is, one does not depend on the evaluation of another, it is immaterial which subquery is executed first. The above rule states that the sum of the cost in executing subqueries is invariant to the order of the subqueries. However, if the *or* expression is part of a conjuncted expression, say $and(Q_{p1}, ..., Q_{pm}, or(Q_1, ...,Q_n))$, then the queries in the disjunction may use the partial result of the conjuncted ancestor queries $(Q_{p1}, ..., Q_{pm})$. The order of execution of the *or* query expression is important with respect to queries $Q_{p1}, ...,$ and Q_{pm}, so the whole disjunction is treated as a single entity. The following illustrates some of the subquery sequences of the example shown in Figure 5.7:

$$[Q_1^a, Q_2^a, Q_3^a, Q_4^s, (Q_5^s, Q_6^a)]$$
$$[Q_1^a, Q_2^a, Q_3^a, (Q_5^s, Q_6^a), Q_4^s]$$
$$[Q_1^a, Q_2^a, (Q_5^s, Q_6^a), Q_3^a, Q_4^s]$$
$$[Q_1^a, Q_2^a, (Q_5^s, Q_6^a), Q_4^s, Q_3^a]$$
$$...$$
$$[(Q_5^s, Q_6^a), Q_2^a, Q_3^a, Q_4^s, Q_1^a].$$

While subqueries in [...] are conjuncted, subqueries in (...) are disjuncted. ☐

For each *or* node of the strategy tree, the decomposition strategy will produce at most one child node which is an aspatial query although there may be a number of child nodes representing spatial queries.

Heuristic Rule 3. For an expression $and(Q_1(S_1), ...,Q_i^a(S_i), ..., Q_j^a(S_j), ..., Q_k^s(S_k), Q_n(S_n))$, suppose $|S_i| = 1$, and $|S_j| = 1$, and $S_i, S_j \subset S_k$, then at least one of Q_i^a or Q_j^a must be executed before Q_k^s. ☐

Discussion. The argument follows from the spatial alternative evaluation schemes stated in Section 5.5.4.2.1.

5.5.4.2.3 SQP Formulation Algorithm

During the generation of different SQPs, it is necessary to minimize the number of SQPs for which cost estimation is performed. A method for generating SQPs has been proposed by [Er87]. This method generates different plans lexicographically [BKK84, Er87, Sed77] in such a way that all arrangements with common initial subsequences are generated consecutively. In this way, the rejection of next few sequences with the same initial subsequence can be made without having to examine them. The method advocated by Er [Er87] is outlined in Appendix B and is used in our query formulations.

Following the heuristic rule 2, the maximum number of sequences that may be generated is $\left\{ \prod_{\text{all } and \text{ node}} \text{number of children!} \right\}$. To specify the order of execution, the strategy tree may be rearranged such that the depth first left to right traversal produces the desired order. Alternatively, an array that describes the ordering can be used. For the example in Figure 5.7, one of the sequences may be represented as follows:

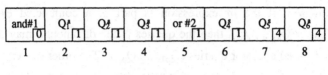

SQP: [Q_1, Q_2, Q_3, (Q_5, Q_6), Q_4].

Figure 5.8 SQP array

An entry consists of a query or a logical operator and the parent index (the inner box). The first entry is always a logical operator and is followed by the children. The SQP can be obtained by recursively scanning the array from left to right. Start from the first entry, its child entries are searched from left to right. When a child entry is a logical operator, the process is repeated from that point. For the array in Figure 5.8, on reaching entry 4, the search for entries 6 and 7 starts. Searching for the remaining children of the 1st entry resumes when the searching of the children of 4th entry is completed.

The SQP fomulating algorithm that uses the lexichographic generator to form different sequences, uses the above array structure to store the current SQP. It selects the best sequence incrementally as outlined below:

Algorithm Formulate

Traverse the strategy tree in breadth first left to right order;
Store each node in a slot in the SQP array;
Generate the SQPs in lexichography order;
For each SQP
    Check each SQP against the heuristic rules
    (make use of the information of previous SQP whenever possible)
    Estimate the cost;
    Retain the best or cheapest SQP;

The algorithm is described in detail in Appendix B.

5.5.4.3 Cost Estimation

The query decomposition strategy discussed so far, in general, provides a means for separating spatial and aspatial predicates such that an existing SQL backend can be used to evaluate aspatial predicates. In order to choose a near optimal subqueries sequence, cost estimation is required. Three crucial points to note about the cost estimation are:

- there is no general agreement [JaK84] on a cost model; the cost estimation is indeed a difficult problem [Chr81].
- the accuracy of the cost estimation depends on the amount of statistical information maintained in the meta-database.
- the estimates for the sizes of intermediate temporary results will tend to differ from the actual figures.

Further, a complicated cost model is expensive to run. Indeed, a recent study [SSD88] showed that the cost of running the optimizer may sometimes overshadow the benefits obtained. A trade off between the optimization effort and execution cost must be taken to minimize the total evaluation cost (execution cost + optimization cost), as illustrated in Figure 5.9.

The cost of processing query Q in the form of its equivalent subquery sequence is the sum of the cost processing each individual subquery. To minimize the cost of the query, it is necessary to choose the subquery sequence which incurs least cost. Let the minimum cost of processing Q(S) be C(Q(S)). Then

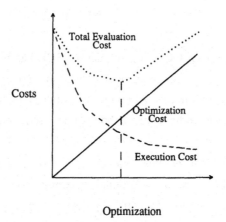

Figure 5.9 Optimization cost

$$C(Q(S)) \equiv \min_{\text{all k sequences}} \left\{ \sum_{i=1}^{N_k} C(Q_i(S_i)) \right\}$$

where N_k is the number of subqueries in kth subquery sequence, and $C(Q_i(S_i))$ is the cost of processing ith subquery in kth sequence.

The proposed extended optimization model can utilize the cost model of the SQL backend to estimate SQL query costs. Making use of the SQL backend cost model however undesirably makes the extended optimizer dependent on a particular backend. In addition, there will be an increase in the overall evaluation cost due to the fact that the execution strategy of each SQL query is practically examined twice. Instead of using the cost model employed by the SQL backend, an empirical parametric cost model would be sufficient to provide a reasonable cost estimation. Such a cost model would use a set of queries, which are deemed to be a representative set, to estimate the execution for different classes of queries, and the cost estimation should involve only a small number of parameters. The cost of the subquery $C(Q_i(S_i))$ is dependent on the type of Q_i (e.g. queries with pure selections) and the cardinality of S_i and the availability of indexes for S_i. This gives a fairly crude estimation. Should a more accurate cost estimation be necessary, the cost function may be tuned as more queries are processed. However, since we are trying to optimize the cost globally, it may be sufficient to use a simple cost estimation model which could be substituted with a finer estimation model if so required. Finding an estimation model itself is a hard problem [Chr81, Ric81], therefore, it is not within the scope of this thesis to give any precise new estimation model. Rather, we assume the use of an existing estimation model, which may be simplistic but relatively useful as in [WoY76].

5.6 Conclusions

An optimization model for GEOQL-like augmented SQL has been proposed. Many of the techniques employed by the proposed extended optimizer are based on existing methods used in current optimizers. Although, the strategy is proposed for a particular extension of SQL, we believe the same method can be used for other languages that are based on SQL.

The major contribution of this chapter is a global optimization strategy for extensions of SQL which require additional indexing structures to materialize the additional relationships. The proposed strategy does not require *extensive modifications* of existing DBMSs. The optimization strategy consists of the following modules: extended logical transformation, decomposition, the generation of subquery sequences, and the selection of the best subquery sequence. Queries are transformed into logically equivalent queries that are more efficient to evaluate. As the SQL backend is not capable of evaluating the spatial components of the extended language, the decomposition breaks a query into several subqueries so that the SQL backend can be used for SQL (UA) subqueries, and a spatial processor, a subsystem for evaluating spatial selection criteria, processes spatial subqueries. With an unordered set of subqueries, all plausible different arrangements of subqueries are considered. This involves the generation of plans and the heuristic pruning of the search space so that only plausible plans are examined.

While such optimizer may not produce the best strategy, it can however produce a reasonable strategy. The approach is simple. It will not involve a substantial increase in the costs. The extension to the DBMS is minor: the SQL parser is modified to parse a larger set of predicates and the extended optimizer is built on top of the existing optimizer.

Chapter 6

Implementation and Experiments

6.1 Justification

To check the correctness of the optimization strategy proposed in Chapter 5 and to justify its feasibility in terms of implementation, an experimental system was built on top of an educational version of SQL called ISQL (Interactive SQL) [McD88]. The ISQL provides an interface to a relational DBMS called the Relational Test Bed (RTB) [McD86]. The RTB is a test bed designed to allow fast implementation of small research databases and the SQL frontend is a query retrieval system designed for educational purposes. Efficiency consideration and the practicability of the systems for large databases were totally ignored in the design of ISQL and RTB [McD86]. There is no support for sophisticated facilities such as buffer management, indexes and an intelligent optimizer.

An empirical comparison with other optimization strategies is difficult since we know of no similar optimization strategies. The main purpose of the implementation is to verify the correctness and feasibility of the proposed optimization model. The implementation was relatively straightforward.

6.2 RTB and ISQL

The RTB provides the basic utility routines, like relational algebraic operations, sorting, and catalog services, from which various small relational database systems may be constructed.

Central to the operations of RTB is the *catalog* which describes the object schema of the database. The catalog consists of a set of tables describing the relevant characteristics of the attribute domains, attributes and relations. Two types of table exist: *domain* tables and *relation view* tables. The domain table stores the information on each of the underlying attribute domains defined in the database and the relation view table stores information on all the relations and user defined view or temporary relations currently active in the database. For each relation defined, an *attribute table* is attached to it. The attribute table describes the position of each attribute field in the table, its name, its domain, and its length. The relation structure supported in the RTB is illustrated in Figure 6.1.

In general, the RTB provides routines to reload and save relations from and to secondary files, and all algebraic operations. There are no indexes, paging capabilities, and buffer management.

The ISQL is constructed on top of the RTB. Without indexes, a join is implemented as a cross product followed by a selection based on the join condition. Recently, the unnesting routines proposed in [GaW87, Kim82] have been included as a part of the logical transformation.

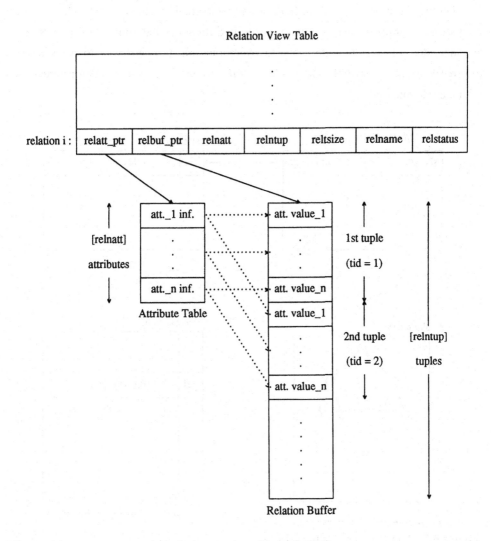

Figure 6.1 Data structure for relations and attributes in RTB

6.3 The GEOQL Database

Associated with each spatial relation are its spatial index (an skd-tree) and an external file describing its spatial attributes. The structure of the database is illustrated in Figure 6.2.

Rather than storing an object MBR with the aspatial attribute data, the object MBR is stored in the external file in order to support the maintenance of the *object data file* as discussed in Chapter 3. In the initial implementation, MBRs were stored as an "abstract data type" (a string representing several fields belonging to the same object) [SRG83].

During the processing of an SQP, a temporary table created as a result of a cross product of multiple spatial relations makes it difficult to determine the ownership of external files. The step undertaken to overcome this problem was to associate the external file and the spatial index with the *gid* attribute of a spatial relation rather than with the relation.

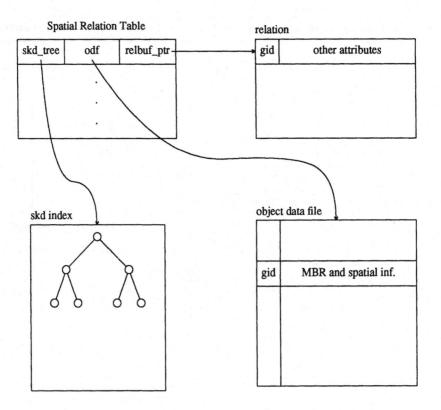

Figure 6.2 Data structure for the GIS

6.4 The Experimental System

The ISQL parser was extended to parse GEOQL queries. The parse tree, a multi-way tree, is implemented as a leftmost-child right-sibling structure [AHU83, p88]. During the parsing, a predicate of the form R_1 *geo_op* R_2 is expanded to $R_1.gid$ *geo_op* $R_2.gid$ to ease the task of query rewriting.

The unnesting routine [Kim82] was extended to unnest GEOQL queries. GEOQL does not introduce any predicate with subqueries as operands; the extension is simple. Consider the following nested query.

```
Query: Find all regions which do not have the same name
       as any roads that intersect railway(s) contained
       in the region.
```

```
SELECT   region.id
FROM     region, road
WHERE    region.name ≠ (
                 select  road.name
                 from    road, railway
                 where   road intersects railway and
                         region contains railway).
```

The unnested equivalent form of the query is as follows:

```
SELECT   region.id
FROM     region, road, railway
WHERE    road intersects railway and
         region contains railway and
         region.name ≠  road.name.
```

It can be noticed that the spatial predicates are unchanged but merely moved to the outer block.

A spatial subsystem was implemented to evaluate a subset of GEOQL spatial predicates. This subsystem accepts one syntactically correct US (unnested spatial) query at a time. Each operand is examined to determine if it is a temporary relation or a base relation. If it is a base relation, then it is checked for any associated indexes. For a given predicate, R_1 geo_op R_2, the following two evaluation strategies are used.

(a)   *[EVALUATION WITHOUT INDEXES]*

   for each tuple t_1 of the relation R_1 do

      get the corresponding spatial object o_1 from the object data file;

      for each tuple t_2 of the relation R_2 do

         get the corresponding spatial object o_2;

         perform spatial evaluation geo_op(o_1, o_2);

         if true then

            append the concatenation of t_1 and t_2 to the resultant relation;

(b)   *[EVALUATION WITH INDEXES]*

   for each tuple t_1 of the relation R_1 do

      get the corresponding spatial object o_1 of t_1;

      search the associated skd-tree of R_2 for all objects o for which

      geo_op(o, o_1) is evaluated true;

      append $t_1 \times \{t_2$: geo_op(o_2, o_1) is true$\}$ to the resultant relation;

The first strategy does not use any spatial indexes, while the second uses the skd-tree to search for all spatial objects of the inner relation that qualify the spatial predicate.

The result of the evaluation is a temporary relation which is a subset of $R_1 \times R_2$.

Given a query, the decomposition routine checks if it contains any spatial predicates. If it does, the query is broken down into subqueries. The decomposition strategy returns a strategy tree to the main calling routine which calls the formulation routine if the strategy tree is not empty. Both algorithms are described in Appendix B and the data structures used to generate different sequences is the SQP array described in Chapter 5.

In the proposed query evaluation strategy, we allow the existence of several instances of a relation. These instances may not have the same number of columns. To maintain the correct relationships, we use the attribute list to link all instances of an attribute. Each attribute instance can be identified via the original attribute and different instances of a relation can also be found using the attribute list. The structure is illustrated in Figure 6.3. With such structure, the query rewriting process becomes straightforward.

We have shown in Chapter 5 that the proposed optimization model produces a correct answer for all the generated evaluation plans it chooses. In order to check that our implementation generated correct SQPs, a small spatial database was built. Queries that involve all the result rewriting rules were used for the tests. The results obtained by different SQPs were found to be exactly the same.

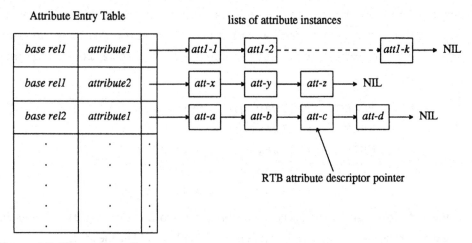

Figure 6.3 The attribute list

In the experimental database, the relation associated with a spatial entity has the attribute *gid* as the primary key. It has an associated object data file and sometimes an skd-tree index.

The RTB provides facilities to create, reload and save relations. The object data files used in this experiment contain only MBRs and gids and are therefore stored as relations for convenience. As such, an object data file can be retrieved as a conventional relation.

We illustrate the features of the implementation using an extended example.

Below is the schema describing a GIS database. Note the GIS is built on top of the existing experimental DBMS, other databases therefore exist. In the following, the relations with postfix "*spatial*" are the object data files.

```
Region          (gid, name, a1)
Lake            (gid, name, a2)
Road            (gid, name, a3)
Railway         (gid, name, a4)
Regionspatial   (gid, mbr)
Lakespatial     (gid, mbr)
Roadspatial     (gid, mbr)
Railwayspatial  (gid, mbr)
```

The example database is described in Appendix C. To illustrate that all SQPs generated for a query produce the same answer, let us consider the following query.

Query: Find all roads and regions that intersect and are such
that the value for attribute a4 of the railway is equal
to the value of attribute a3 of some road for which
road.a3 > 20.

```
SELECT   railway.gid, region.gid
FROM     railway, region, road
WHERE    railway.a4 = road.a3 and
         road.a3 > 20 and
         railway intersects region.
```

For the given data (see Appendix C), the answer to the above query is given in
Figure 6.4. The answers produced by all generated SQPs should be the same.

gid	gid
1004	0004
1004	0007
1004	0019

Figure 6.4 The answer

Two subqueries formed by the decomposition algorithm are given below:

Q_1:
```
SELECT   *
FROM     railway, region
WHERE    railway intersects region.
```

Q_2:
```
SELECT   *
FROM     railway, road
WHERE    railway.a4 = road.a3 and
         road.a3 > 20.
```

For the above example, the decomposition algorithm is able to detect that the join
railway.a4 = road.a3 is in effect a restriction due to the fact that the relation road does
not participate in other multiple relation predicates or in the target list. This sort of
predicate is known as an existential restriction [Dat86]. Therefore, the predicates
railway.a4 = road.a3 and *road.a3 > 20* are grouped together to form a subquery. On the
other hand, if the relation road participates in the target list or in other spatial predicate(s)

(e.g. *railway.a4 = road.a3 and road.a3 > 20 and road intersects railway*) then the aspatial predicates would need to be decomposed into two separate groups. It is obvious that with above decomposition, the flexibility of using a spatial indexing structure is supported. Should it be considered necessary to evaluate the join before the spatial predicate, the two subqueries (one that contains the restriction and one that contains the join) should be rewritten as one subquery to allow the SQL backend to evaluate the aspatial predicates with more freedom.

The subquery tree for the example query consists of an *and* node with the two subqueries being the children. The SQP array contains these three nodes, with the *and* node being the first element. Using the SQP array, the following two subquery sequences are generated.

$$[Q_1, Q_2]$$
$$[Q_2, Q_1]$$

Two different ways of merging the partial results using the result rewriting rule 1 will be used to show how each SQP may be executed. For illustration, all attributes are selected during the processing and projections are performed only when the answer is required.

During the processing of a subquery sequence, a temporary relation with two attributes (originate from two different base relations) having the same name may be resulted. In order to avoid the confusion as to which relation the attribute originally belongs, we use T.R.att to denote an attribute *att* of base relation R in the temporary relation T.

Firstly, consider the evaluation of Q_1 and then Q_2 using the partial result produced by Q_1.

SELECT   *
INTO   Tmp00001
FROM   railway, region
WHERE   railway intersects region.

railway			region		
gid	name	a4	gid	name	a1
1001	railway1	01	0007	region7	75
1001	railway1	01	0001	region1	11
1001	railway1	01	0004	region4	22
1001	railway1	01	0002	region2	11
1001	railway1	01	0019	rusden	42
1002	railway2	11	0001	region1	11
1002	railway2	11	0002	region2	11
1002	railway2	11	0019	rusden	42
1004	railway4	22	0007	region7	75
1004	railway4	22	0004	region4	22
1004	railway4	22	0019	rusden	42
1005	railway5	04	0005	region5	34

Figure 6.5a Execution of the spatial subquery Q_1

SELECT   *
INTO   Tmp00002
FROM   Tmp00001, road
WHERE   (Tmp00001.a4 = road.a3 and
              road.a3 > 20).

railway			region			road		
gid	name	a4	gid	name	a1	gid	name	a3
1004	railway4	22	0004	region4	22	1014	road4	22
1004	railway4	22	0007	region7	75	1014	road4	22
1004	railway4	22	0019	rusden	42	1014	road4	22

Figure 6.5b Restriction using partial result of the spatial subquery

Figure 6.5b shows the partial result of the restriction using the partial result from the evaluation of the spatial subquery. The answer is a projection on the result of the restriction.

The following shows the independent execution of the restriction query (Q_2) and in this case the answer is the join (and then projection) between the result of Figures 6.5a and 6.6 base on the attributes of the relation railway.

```
SELECT    *
INTO      Tmp00003
FROM      railway, road
WHERE     (railway.a4 = road.a3 and
          road.a3 > 20).
```

railway			road		
gid	name	a4	gid	name	a3
1004	railway4	22	1014	road4	22

Figure 6.6 Result of execution of aspatial subquery Q_2

The following is the query that merges the partial results. Note that the attribute renaming is handled with the *attribute entry table*. Notice that a join on the common attributes that have involved in the evaluation (e.g. railway.a4 and railway.gid) will also give the same answer. The extension requires extra book keeping routines and our implementation can be easily extended to handle such cases.

```
SELECT    *
FROM      TMP00001, TMP00003
WHERE     TMP00001.railway.gid = TMP00003.railway.gid and
          TMP00001.railway.name = TMP00003.railway.name and
          TMP00001.railway.a4 = TMP00003.railway.a4.
```

Now consider the second ordering, $[Q_2, Q_1]$. Since the independent evaluation of Q_2 and Q_1 is the same as the previous strategy, we shall only show how Q_1 makes use of the result of Q_2.

```
SELECT    *
INTO      Tmp00005
FROM      Tmp00003, region
WHERE     Tmp00003 intersects region.
```

railway			road			region		
gid	name	a4	gid	name	a3	gid	name	a1
1004	railway4	22	1014	road4	22	0004	region4	22
1004	railway4	22	1014	road4	22	0007	region7	75
1004	railway4	22	1014	road4	22	0019	rusden	42

Figure 6.7 Execution of spatial subquery using (Q_1) partial result

The relation obtained as a result of executing the spatial subquery using the partial result formed by the restriction is shown in Figure 6.7. The answer is again a simple projection of the result.

Examples involving the other two rewriting rules were formulated and tested.

Without indexes and optimization strategies in RTB, we had difficulty in performing any empirical analysis. It should be noted that other than calling appropriate RTB/ISQL routines, we did not alter any supported routines.

6.5 Discussion

6.5.1 Extensibility

The extended optimization approach, although presented as a technique to be built on top of an existing backend, forms a basis for optimizing queries of any complexity — general queries. In the place of subqueries created by our decomposition algorithm, the predicates that form the qualifications of subqueries can be transformed into a sequence of relational algebra operations or a high level language program [Fre87b]. Then the global optimization follows.

The usefulness of heuristic rules greatly depends on the nature of the operators that are supported. Therefore only those rules applicable to general queries should be supported and the number of rules should be kept as small as possible. Then only a new operator can be added to the extended DBMS without having to make substantial alteration of the system.

6.5.2 Limitations

The major limitation with the extended optimization model is that we must be able to access the existing DBMS at other than the user's level. The UA subqueries formed by the decomposition strategy are syntactically and semantically correct. It would be very inefficient to submit each query to the DBMS and then make use of the temporary results as part of the execution strategy. We need to able to control the buffer management and to access the temporary storage without having to submit the decomposed UA queries at the user's level. SQL queries with INTO clauses create more than just the problem of creating temporary tables, but they also involve expensive rereading operations.

The separate processing of spatial and aspatial predicates entails communication between the two processors. This may cause a bottleneck communication problem and degrade the query processing. More study in this area is required.

6.6 Conclusion

In this chapter, we have shown that the optimization strategy proposed in Chapter 5 is feasible. It is obvious that the optimization strategy enhances the query processing efficiency of the DBMS backend. However, the extent to which the extended optimizer can support efficient query processing remains to be quantified and this study is only possible with a more sophisticated SQL backend which has buffer management and a conventional optimization strategy.

Chapter 7

Conclusions

The main contributions of this thesis are listed below:

(1) The identification of the requirements that a GIS should satisfy and a general review of existing interface languages and existing GISs.

(2) A detailed study of existing spatial indexing structures that provide access to spatial objects based on proximity and a review of fundamental optimization strategies.

(3) The proposal of *an augmented external interface query language* called GEOQL. GEOQL is an extension of the relational query language SQL and provides additional operators for the support of proximity queries.

(4) The proposal and analysis of an efficient spatial indexing structure called the *skd-tree*.

(5) The proposal of an implementation model and architecture for extending an existing DBMS to support the functionality of a GIS.

(6) The proposal of *an extended optimization strategy* for processing queries that contain both spatial and aspatial predicates and which effectively utilizes a DBMS backend for the evaluation of the spatial components.

The review of previous work gives an insight to the problems associated with geographic information processing. It provides guidelines for any new proposals regarding the construction of a GIS.

A query language for a GIS should be both powerful and easy to use. GEOQL is an extension of SQL that allows selection based on both spatial and aspatial criteria.

Important to the efficiency of query processing is the underlying indexing structures. Unlike the objects stored in conventional databases, the objects in a GIS have spatial characteristics, and the retrieval of an object based on its location or its proximity to other objects must be supported. An extensive review of recently proposed indexing structures has been presented.

The existing kd-trees are efficient indexing structures for multi-dimensional point data, however they are not suitable for indexing non-zero sized spatial objects. A new data structure that is based on the kd-tree and is suitable for indexing non-zero sized objects is proposed in this thesis. The empirical results that are presented show that the skd-tree is indeed an efficient structure compared to other widely used indexing structures [Gut84, MHN84, Ros85]. The skd-tree provides direct support for *containment search*, as well as *intersection search*. It also supports searches using a set of query objects, making the evaluation of spatial predicates more efficient.

An architecture for a GIS that utilizes a relational DBMS has been proposed (see Figure 7.1). An optimization model for evaluating GEOQL queries has been presented. The steps involved in the optimization of GEOQL queries include logical transformation, query decomposition, subquery plan formulation and the best subquery plan selection. The strategies proposed for query optimization are designed to make maximum use of the spatial indexing structures as well as utilizing the query optimizer within the relational backend.

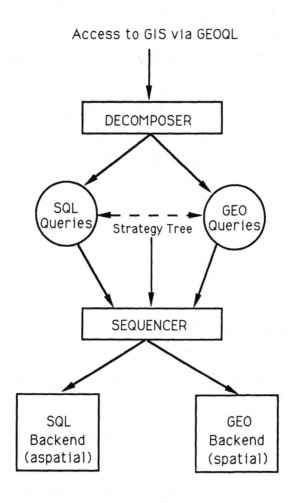

Figure 7.1 The proposed GIS architecture

The proposed query optimization model can be generalized to be a basis for other extensions of a relational DBMS.

Areas of future research include the production of accurate cost models for ordering the subqueries during the optimization process. The spatial properties of the entities in a GIS provide an extra dimension to this problem compared to cost estimation in conventional DBMSs. Other issues for future research include concurrency control and integrity constraint checking in an extended DBMS environment.

To summarize, this thesis has provided a foundation for extending a DBMS for geographical applications. The main motivations for the design are:

(1) the efficient answering of queries that involve both spatial and aspatial operators and

(2) the use of a relational backend in order to minimize the implementation effort required to develop a GIS.

We have demonstrated the feasibility of this approach in this thesis.

References

[AbS83]   D. J. Abel, J. L. Smith: A data structure and algorithm based on a linear key for a rectangle retrieval problem. *International Journal of Comp. Vision, Graphics, and Image Processing 24*, 1, 1-13 (1983).

[AbS84]   D. J. Abel, J. L. Smith: A data structure and query algorithm for a database of areal entities. *Aust. Comp.J. 16*, 4, 147-154 (1984).

[AbS86]   D. J. Abel, J. L. Smith: A relational GIS database accommodating independent partitionings of the region. *Proc. 2nd International Symposium on Spatial Data Handling*, Seattle, WA, 213-224 (1986).

[AbS87]   D. J. Abel, J. L. Smith: A kernel-shell approach to an extended relational spatial database management system. Unpublished paper, CSIRO (1987).

[Abe84]   D. J. Abel: A B+ structure for large quadtrees. *International Journal of Comp. Vision, Graphics, and Image Processing 27*, 1, 19-31 (1984).

[Abe86]   D. J. Abel: Bit-interleaved keys as the basis for spatial access in a front-end spatial database management system. In: B. Diaz, S. Bell, (eds): *Spatial Data Processing Using Tesseral Methods*, Natural Environment Research Council, 163-177 (1986).

[ARC90]   *ARC News 12*, 1, Enivironmental Systems Research Institute, (1990).

[AHU83]   A. V. Aho, J. E. Hopcroft, J. K. Ullman: *Data structures and algorithms*. Reading, MA: Addison-Wesley 1983.

[ABD80]   F. Antonacci, L. Bartolo, P. Dell'Orco, V. Spadavecchia: AQL - a relational data base management system and its geographical applications. In: *Data base techniques for pictorial applications*, Berlin-Heidelberg-New York: Springer-Verlag, 569-599 (1980).

[ABC76]   M. M. Astrahan, M. W. Blasgen, D. D. Chamberlin, K. P. Eswaran, J. N. Gray, P. P. Griffiths, W. F. King, R. A. Lorie, P. R. McJones, J. W. Mehl, G. R. Putzolu, I. L. Traiger, B. W. Wade, V. Watson: System R: relational approach to database management. *ACM Trans. on Database Sys. 1*, 2, 97-137 (1976).

[BaR86]   F. Bancilhon, R. Ramakrishnan: An amateur's introduction to recursive query processing strategies. *Proc. ACM SIGMOD International Conf. on Management of Data*, Washington, D.C., 16-52 (1986).

182

[BaK86]   J. Banerjee, W. Kim: Supporting VLSI geometry operations in a database system. *IEEE Proc. Data Engineering Conf.*, Los Angeles, CA, 409-415 (1986).

[BaB81]   R. Barrera, A. Buchmann: Schema definition and query language for a geographical data base system. *Comp. Architecture for Pattern Analysis and Image Database Management*, 250-256 (1981).

[BaM86]   D. S. Batory, M. Mannino: Panel on extensible database systems. *Proc. ACM SIGMOD Int. Conf. on Management of Data*, Washington, D.C., 187-190 (1986).

[Bat86]   D. Batory : Extensible cost models and query optimization in genesis. In: G. Lohman(ed): *A quarterly bulletin of the Computer Society of the IEEE technical committee on Database Engineering 9*, 4, 30-36 (1986).

[BaM72]   R. Bayer, E. McCreight: Organization and maintenance of large ordered indices. *Acta Informatica 1*, 3, 173-189 (1972).

[BER85a]   D. A. Beckley, M. W. Evens, V. K. Raman: Empirical comparison of associative file structures. *Proc. International Conf. on Foundations of Data Organisation*, Kyoto, Japan, 315-319 (1985).

[BER85b]   D. A. Beckley, M. W. Evens, V. K. Raman: An experiment with balanced and unbalanced K-D trees for associative retrieval. *Proc. IEEE COMPSAC - Comp. Software & Applications Conf.*, Chicago, IL, 256-262 (1985).

[BER85c]   D. A. Beckley, M. W. Evens, V. K. Raman: Multikey retrieval from K-D trees and quad trees. *Proc. ACM SIGMOD International Conf. on Management of Data*, Austin, Texas, 291-301 (1985).

[Ben75]   J. L. Bentley: Multidimensional binary search trees used for associative searching. *Comm. ACM 18*, 9, 509-517 (1975).

[BeF79]   J. L. Bentley, J. H. Friedman: Data structures for range searching. *ACM Comp. Surveys 11*, 4, 397-409 (1979).

[Ben79a]   J. L. Bentley: Multidimensional binary search trees in database applications. *IEEE Trans. on Software Eng. SE-5*, 4, 333-340 (1979).

[Ben79b]   J. L. Bentley: Decomposable searching problems. *Information Processing Letters 8*, 5, 244-251 (1979).

[BeS77]   R. Berman, M. Stonebraker: GEO-QUEL: A system for the manipulation and display of geographic data, *Comp. Graphics 11*, 2, 186-191 (1977).

[BeC81]   P. A. Bernstein, D. M. Chiu: Using semi-joins to solve relational queries. *Journal of ACM 28*, 25-40 (1981).

[BGW81]   P. A. Bernstein, N. Goodman, E. Wong, C. L. Reeve, J. B. J. Rothnie: Query processing in a system for distributed databases (SDD-1). *ACM Trans. on Database Sys. 6*, 4, 602-625 (1981).

[BlE77]   M. W. Blasgen, K. P. Eswaran: Storage and access in relational databases. *IBM Sys. J. 16*, 4, 363-377 (1977).

[Bra84]   K. Bratbergsengen: Hashing methods and relational algebra operations. *Proc. 10th International Conf. on Very Large Data Bases*, Singapore, 323-333 (1984).

[Bur80]   J. F. C. Burgueno: A geographical data base. In: *Data base techniques for pictorial applications.* Berlin-Heidelberg-New York: Springer-Verlag, 347-363 (1980).

[Bur83]   W. A. Burkhard: Interpolation-based index maintenance. *BIT 23*, 274-294 (1983).

[BKK84]   F. W. Burton, V. J. Kollias, J. G. Kollias: Permutation backtracking in lexicographic order. *The Computer Journal 27*, 4, 373-376 (1984).

[CeS82]   F. Cesarini, G. Soda: Binary trees paging, *Information Systems 7*, 4, 337-344 (1982).

[Cha74]   D. D. Chamberlin: SEQUEL : A structured English query language, *Proc. ACM SIGMOD Workshop on Data Description, Access and Control 1*, Ann Arbor, Michigan, 249-264 (1974).

[CAE76]   D. D. Chamberlin, M. M. Astrahan, K. P. Eswaran, P. P. Griffiths, R. A. Lorie, J. W. Mehl, P. Reisner, B. W. Wade: SEQUEL 2: A unified approach to data definition, manipulation and control. *IBM J. of Research and Development 20*, 6, 560-575 (1976).

[Cha77]   H. Chang: Bubbles for relational database. Report RC 7485, IBM Thomas J. Watson Research Center, Yorktown Heights, New York (1977).

[ChF79]   J. M. Chang, K. S. Fu: Extended K-D tree database organization: A dynamic multi-attribute clustering method. *Proc. IEEE COMPSAC - Comp. Software & Applications Conf.*, Chicago, IL, 39-43 (1979).

[CLW80]   S. K. Chang, B. S. Lin, R. Walser: A generalized zooming technique for pictorial database systems. In: S. K. Chang, K. S. Fu (eds): *Pictorial Information Systems*, Springer-Verlag, 257-286 (1980).

[ChF80a]   N. S. Chang, K. S. Fu: Query-by-pictorial-example. *IEEE Trans. on Software Eng. SE-6*, 6, 519-524 (1980).

[ChF80b]   N. S. Chang, K. S. Fu: A relational database system for images. In: S. K. Chang, K. S. Fu (eds): *Pictorial Information Systems*, Springer-Verlag, 288-321 (1980).

[ChF81]   N. S. Chang, K. S. Fu: Picture query language for pictorial data-base systems. *IEEE Computer 14*, 11, 23-33 (1981).

[CLS84]   J. M. Cheng, C. R. Loosley, A. Shibamiya, P. S. Worthington: IBM database 2 performance: Design, implementation and tuning. *IBM Sys. J. 3*, 2, 189-214 (1984).

[CCK81]   M. Chock, A.F. Cardenas, A. Klinger: Manipulating data structures in pictorial information systems. *IEEE Computer 14*, 11, 43-52 (1981).

[CCK81]   M. Chock, A.F. Cardenas, A. Klinger: Database structure and manipulation capabilities of a picture management system (PICDMS). *IEEE Trans. Pattern Anal. Machine Intell. PAMI-6*, 4, 484-492 (1981).

[Chr81]   S. Christodoulakis: Estimating selectivities in data bases. Ph. D. Thesis, Tech. Rep CSRG-136, Computer Systems Research Group, University of Toronto, Canada (1981).

[Cod70]   E. F. Codd: A relational model of data for large shared data banks. *Comm. ACM 13*, 6, 377-387 (1970).

[Cod71a]   E. F. Codd: Relational completeness of data base sublanguages. In: R. Rustin (ed): *Data base systems*, Courant Comp. Science Symposia, vol. 6, Prentice-Hall, Englewood Cliffs, New Jersey, 65-98 (1971).

[Cod71b]   E. F. Codd: A data base sublanguage founded on the Relational calculus. *Proc. ACM SIGFIDET Workshop on data description, access and control*, San Diego, California, 35-68 (1971).

[Cod81]   E. F. Codd: SQL/DS - What it means. *Australiasian Computer world*, 13-15 (1981).

[Com79]   D. Comer: The ubiquitous B-Tree. *ACM Comp. Surveys 11*, 2, 121-137 (1979).

[Dat84a]   C. J. Date: *A guide to DB2*, Reading, MA: Addison-Wesley (1984).

[Dat84b]   C. J. Date: A critique of the SQL database language. *ACM SIGMOD Record 14*, 3, 8-54 (1984).

[Dat86]   C. J. Date: *An introduction to database systems, volume 1, 4th Edition*. Reading, MA: Addison-Wesley (1986).

[Dat87]   C. J. Date: *A guide to INGRES*. Reading, MA: Addison-Wesley (1987).

[DaP85]   W. A. Davis, M. Palat: Data base management for geo-data. Tech. Rep 85-17, Univ. of Alberta, Edmonton, Canada (1985).

[DaH85]   W. A. Davis, C. H. Hwang: File organization schemes for geometric data. Tech. Rep. 85-14, Univ. of Alberta, Edmonton, Canada (1985).

[DKO84]   D. J. DeWitt, R. H. Katz, F. Olken, L. D. Shapiro, M. R. Stonebraker, D. Wood: Implementation techniques for main memory database systems. *Proc. ACM SIGMOD International Conf. on Management of data*, Boston, MA, 1-8 (1984).

[EaZ82]   C. M. Eastman, M. Zemankova: Partially specified nearest neighbour using kd Trees. *Information Processing Letters 15*, 2, 53-56 (1982).

[Er87]   M. C. Er: An efficient implementation of permutation backtracking in lexicographic order. *The Computer Journal 30*, 3, 282 (1987).

[FNP79]   R. Fagin, J. Nievergelt, N. Pippenger, H. R. Strong: Extendible hashing - A fast access method for dynamic files. *ACM Trans. on database sys. 4*, 3, 315-344 (1979).

[FSR87]   C. Faloutsos, T. Sellis, N. Roussopoulos: Analysis of object oriented spatial access methods. *Proc. ACM SIGMOD International Conf. on management of data*, San Francisco, California, 426-439 (1987).

[FiB74]   R. A. Finkel, J. L. Bentley: Quad Trees: A data structure for retrieval on composite keys. *Acta informatica 4*, 1-9 (1974).

[Fra82]   A. Frank: Mapquery: Data base query language for retrieval of geometric data and their graphical representation. *Comp. Graphics 16*, 3, 199-270 (1982).

[Fre87]   M. Freeston: The BANG file: A new kind of grid file. *Proc. ACM SIGMOD International Conf. on management of data*, San Francisco, California, 260-269 (1987).

[FrG86]   J. C. Freytag, N. Goodman: Rule-based translation of relational queries into iterative programs. *Proc. ACM SIGMOD International Conf. on management of data*, Washington, D.C., 206-214 (1986).

[Fre87a] J. C. Freytag: *Translating relational queries into iterative programs.* Lecture Notes in Comp. Science 261, Springer-Verlag, New York (1987).

[Fre87b] J. C. Freytag: A rule-based view of query optimization. *Proc. ACM SIGMOD Int. Conf. on Management of Data*, San Francisco, California, 173-180 (1987).

[FBF77]   J. H. Friedman, J. L. Bentley, R. A. Finkel: An algorithm for finding best matches in logarithmic expected time. *ACM Trans. on Math. Software 3*, 3, 209-226 (1987).

[FKS81]   D. Fussell, Z. M. Kedem, A. Silberschatz: Deadlock removal using partial rollback in database systems. *Proc. ACM SIGMOD International Conf. on management of data*, Ann Arbor, Michigan, 65-73 (1981).

[GaR88]   M. N. Gahegan, S. A. Roberts: Intelligent object-oriented geographic information system. Unpublished paper, Dept. of Computer Studies, Uni. of Leeds, Leeds, England (1988).

[GaW87]   R. A. Ganski, H. K. T. Wong: Optimization of nested SQL queries revisited. *Proc. ACM SIGMOD Int. Conf. on Management of Data*, San Francisco, California, 23-33 (1987).

[Gen90]   *G5/New 2*,1. The Generation 5 Technology, (1990).

[Goh86]   G. K. Goh: Screen oriented query languages. M.Sc. Thesis, Comp. Science Dept., Monash Univ., Melbourne, Australia (1986).

[GoM87]   G. K. Goh, K. J. McDonell: QBE++: A QBE-Like Query Language with Improved Semantics. *Proc. 2nd Pan Pacific Comp. Conf.*, Singapore, 321-334 (1987).

[GoS82]   N. Goodman, O. Shmueli: Tree queries: A simple class of relational queries. *ACM Trans. on Database Sys. 7*, 4, 653-677 (1982).

[Gra86]   G. Graefe: Software modularization with the EXODUS optimizer generator. In: G. Lohman (ed): *A quarterly bulletin of the Computer Society of the IEEE technical committee on Database Engineering 9*, 4, 37-43 (1986).

[GrD87]   G. Graefe, D. J. DeWitt: The EXODUS optimizer generator. *Proc. ACM SIGMOD International Conf. on the Management of Data*, San Francisco, 160-172 (1987).

[GrM81]   J. Grant, J. Minker: Optimization in deductive and conventional database systems. In: H. Gallaire, J. Minker, J. M. Nicolas (eds): *Advances in Database Theory*, Vol. 1, Plenum, 195-234 (1981).

[Gra87]   J. Grant: *Logical introduction to databases*. Harcourt Brace Jovanovich, USA (1987).

[Gun88]   O. Gunther: *Efficient structures for geometric data management*. Lecture Notes in Computer Science 337, Springer-Verlag (1988).

[Gun89]   O. Gunther: The design of the cell tree: an object-oriented index structure for geometric databases. *IEEE 5th Int. Data Engineering Conf.*, 598-605 (1989).

[Gut84]   A. Guttman: R-trees: A dynamic index structure for spatial searching. *Proc. ACM SIGMOD Int. Conf. on Management of Data*, Boston, MA, 47-57 (1984).

[HaJ84]   D. J. Haderle, R. D. Jackson: IBM database 2 overview. *IBM Sys. J. 23*, 2, 112-125 (1984).

[Hae78]   T. Haerder: An implementation technique for a generalized access path structure. *ACM Trans. on Database Sys. 3*, 3, 285-298 (1978).

[Hal74]   P. A. V. Hall: Common subexpression identification in general algebraic systems. *Tech. Rep. UKSC 0060*, IBM UK Scientific Center, Peterlee, England (1974).

[Hal76]   P. A. V. Hall: Optimization of single expressions in a relation data base system. *IBM J. of Research and Development 20*, 3, 244-257 (1976).

[HaZ80]   M. Hammer, S. B. Zdonik Jr.: Knowledge-base query processing. *Proc. 6th Int. Conf. on Very Large Data Bases*, Montreal, 223-233 (1980).

[Har80]   R. M. Haralick: A spatial data structure for geographic information. In: H. Freeman, G. G. Pieroni (eds): *Map Data Processing*, Maratea, Italy, Academic Press, New York, 63-100 (1986).

[Hei63]   W. P. Heising: *IBM Systems Journal 2.* 114-115 (1963).

[HSW]   G. D. Held, M. R. Stonebraker, E. Wong: INGRES - A relational data base management system. *Proc. AFIPS National Comp. Conf. 44*, 409-416 (1975).

[HeY79]   A. Hevner, S. B. Yao: Query processing in distributed database systems. *IEEE Trans. on Software Eng. SE-5*, 177-187 (1979).

[HiN83]   K. Hinrichs, J. Nievergelt: The grid file: A data structure designed to support proximity queries on spatial objects. *Proc. International Workshop on Graphtheoritic Concepts in Comp. Science.* Trauner-Verlag, 100-113 (1983).

[Hin85]   K. Hinrichs: Implementation of the grid file: Design concepts and experience. *BIT 25*, 569-592 (1985).

[Hue80]   G. Huet: Confluent reductions: Abstract properties and applications of term rewriting systems. *J. ACM 27*, 797-821 (1980).

[IBM84]   Special issues on DB2. *IBM Sys, J. 23*, 2 (1984).

[ISO86]   ISO: Information processing systems - Database language SQL. Draft International Standard 9075 (1986).

[IoW87]   Y. E. Ioannidis, E. Wong: Query optimization by simulated annealing. *Proc. ACM SIGMOD International Conf. on Management of Data*, San Francisco, California, 9-22 (1987).

[JaS82]   M. Jarke, J. Schmidt: Query processing strategies in pascal/R relational database management system. *Proc. ACM SIGMOD International Conf. on Management of Data*, Orlando, Florida, 256-264 (1982).

[JaK84]   M. Jarke, J. Koch: Query optimization in database systems. *ACM Comp. Surveys 16*, 2, 111-152 (1984).

[Jar85]   M. Jarke: Common subexpression isolation in multiple query optimization. In: W. Kim, D. S. Reiner, D. S. Batory (eds): *Query Processing in Database Systems*, Springer-Verlag, New York, 191-250 (1985).

[JoC88]   T. Joseph, A. F. Cardenas: PICQUERY: A high level query language for pictorial database management. *IEEE Trans. on Software Eng. 14*, 5, 630-638 (1988).

[KaY83]   Y. Kambayashi, M. Yoshikawa: Query processing utilizing dependencies and horizontal decomposition. *Proc. ACM SIGMOD International Conf. on Management of Data*, 55-67 (1983).

[KeW87]   A. Kemper, M. Wallrath: An analysis of geometric modeling in database systems. *ACM Comp. Surveys 19*, 1, 47-91 (1987).

[Kim82]   W. Kim: On optimizing an SQL-like nested-query. *ACM Trans. on Database Sys. 7*, 3, 443-469 (1982).

[Kim85]   W. Kim: Global optimization of relational queries: A first step. In: W. Kim, D. S. Reiner, D. S. Batory (eds): *Query Processing in Database Systems*, Springer-Verlag, New York, 206-216 (1985).

[KBC88]   W. Kim, N. Ballou, H.-T. Chou, J. F. Garza, D. Woelk: Features of the ORION object-oriented database system. In: W. Kim, F. H. Lochovsky (eds): *Object-Oriented Concepts, Databases, and Applications*, Addison-Wesley, 251-282 (1988).

[Kin81]   J. J. King: QUIST: A system for semantic query optimization in relational databases. *Proc. 7th Int. Conf. on Very Large Data Bases*, Cannes, France, 510-517 (1981).

[KTM83]   M. Kitsuregawa, H. Tanaka, T. Moto-Oka: Application of hash to data base machine and its architecture. *New Generation Computing*, 62-74 (1983).

[Kli71]   A. Klinger: Patterns and search statistics. In: J. S. Rustagi (ed): *Optimizing Methods in Statistics*, Academic Press, New York (1971).

[Klu82]   A. Klug: Access paths in the ABE statistical query facility. *Proc. ACM SIGMOD International Conf. on Management of Data*, Orlando, Florida, 161-173 (1982).

[Knu]   D. E. Knuth: *Fundamental Algorithms, 2nd Edition*. The art of computer programming, vol. 1, Reading, MA: Addison-Wesley (1973).

[Kri84]   H. P. Kriegel: Performance comparison of index structures for multi-key retrieval. *Proc. ACM SIGMOD International Conf. on Management of Data*, Boston, MA, 186-196 (1984).

[KrS86]   H. Kriegel, B. Seeger: Multidimensional order preserving linear hashing with partial expansion. *Proc. International Conf. on Database Theory*, Rome, Italy, Springer-Verlag, New York, 203-220 (1986).

[KrS88]   H. Kriegel, B. Seeger: PLOP-Hashing: A grid file without directory. *IEEE 4th International Conf. on Data Engineering*, L.A., California, 369-376 (1988).

[Lar78]   P. Larson: Dynamic hashing. *BIT 13*, 184-201 (1978).

[LeW77]   D. T. Lee, C. K. Wong: Worst-Case analysis for region and partial region searches in multidimensional binary search trees and balanced quad trees. *Acta Informatica 9*, 1, 23-29 (1977).

[LeC86]   T. J. Lehman, M. J. Carey: Query processing in main memory database management systems. *Proc. SIGMOD International Conf. on the Management of Data*, Washington, D.C., 239-250 (1986).

[Leh86]   T. J. Lehman: Design and performance evaluation of a main memory relational database system. Tech. Rep. #656, Ph.D. Thesis, Comp. Science Dept., Univ. of Wisconsin, Madison, Wisconsin (1986).

[LiC79]   B. S. Lin, S. K. Chang: Picture algebra for interface with pictorial database systems. *Proc. IEEE COMPSAC - Comp. Software & Applications Conf.*, Chicago, IL, 525-530 (1979).

[LMP87]   B. Lindsay, J. McPherson, H. Pirahesh: A data management extension architecture. *Proc. ACM SIGMOD International Conf. on the Management of Data*, San Francisco, California, 220-226 (1987).

[Loh87]   G. Lohman: Grammar-like functional rules for representing query optimization alternatives. RJ 5992, IBMTJ, San Jose, CA (1987).

[LNN90]   H.-J. Lu, A.D. Narasimhalu, A. Ngu, B. C. Ooi, H. H. Pang, R. Price, L. S. Wong,J. J. Lim: MOODS — A multimedia object-oriented database system. Technical Report, Institute of Systems Science (1990).

[MaL86]   L. Mackert, G. Lohman: R* optimizer validation and performance evaluation for local query. *Proc. ACM SIGMOD International Conf. on Management of Data*, Washington, D.C., 84-95 (1986).

[MSO86]   D. Maier, J. Stein, A. Otis, A. Purdy: Development of an object-oriented DBMS. Proc. OOPSLA 86 Object-Oriented Programming Systems, Languages and Applications, Portland, 472- 481 (1986).

[MaC80]   P. E. Mantey, E. D. Carlson: Integrated geographic data bases: the GADS experience. In: *Data Base Techniques for Pictorial Applications*, Springer-Verlag, 173-198 (1980).

[Mar82]   J. J. Martin: Organisation of geographic data with quad trees and least square approximation. *Proc. Conf. on Pattern Recognition and Image Processing*, Las Vegas, Nevada, 458-463 (1982).

[MHN84]   T. Matsuyama, L.V. Hao, M. Nagao: A file organization for geographic information systems based on spatial proximity. *Int. Journal Comp. Vision, Graphics, and Image Processing 26*, 3, 303-318 (1984).

[McD86]   K. J. McDonell: An overview of the relational test bed (RTB). Tech. Rep. 81, Dept. of Comp. Science, Monash University, Australia (1986).

[McD88]   K. J. McDonell: Instructional SQL (ISQL) user's guide. Tech. Rep. 101, Dept. of Comp.Science, Monash University, Australia (1988).

[McK84]   D. M. McKeown: Digital cartography and photo interpretation from a data base viewpoint. In: G. Gardarin and E. Gelenbe (eds): *New Applications of Data Bases*, Academic Press, London, 19-42 (1984).

[McK83]   D. M. McKeown Jr: MAPS: the organization of a spatial data base system using imagery, terrain and map data. Tech. Rep. CMU-Comp. Science-83-136, Comp. Sc. Dept., Carnegie-Mellon University (1983).

[Mei82]   A. Meier: A framework for specifying geographic database applications. *Proc. IEEE COMPSAC - Comp. Software & Applications Conf.*, Chicago, IL, 476-481 (1982).

[Mei85]   A. Meier: A graph grammar approach to geographical databases. *Information Systems 10*, 9-19 (1985).

[Mis82]   M. Missikoff: A domain based internal schema for relational database machines. *Proc. ACM SIGMOD International Conf. on Management of Data*, Orlando, Florida, 215-224 (1982).

[NaW79]   G. Nagy, S. Wagle: Geographic data processing. *ACM Comp. Surveys 11*, 2, 139-181 (1979).

[Ngu89]   A. H. H. Ngu: Conceptual transaction modeling. *IEEE Tran. on Knowledge and Data Engineering 1*, 4, 508-518 (1989).

[NgO90]   A. H. H. Ngu, B. C. Ooi: Data model issues for object-oriented database design, ISS Technical Report, National University of Singapore (1990).

[NHS84]   J. Nievergelt, H. Hinterberger, K. C. Sevcik: The grid file: An adaptable, sysmetric multikey file structure. *ACM Trans. on Database Sys. 9*, 1, 38-71 (1984).

[NiH85]   J. Nievergelt, K. Hinrichs: Storage and access structures for geometric data bases. *International Conf. on Foundations of Data Organization*, Kyoto, Japan, 335-345 (1985).

[OFS84]   J. Ong, D. Fogg, M. Stonebraker: Implementation of data abstraction in the relational database system ingres. *ACM SIGMOD Record 14*, 1, 1-14 (1984).

[OMS87]   B. C. Ooi, K. J. McDonell, R. Sacks-Davis: Spatial kd-tree: An indexing mechanism for spatial databases. *Proc. IEEE Comp. Software & Applications Conf.*, Tokyo, Japan, 433-438 (1987).

[Ooi88]   B. C. Ooi: *Efficient Query Processing in a Geographic Information System.* Ph.D. Thesis, Monash University (1988).

[OoS89]   B. C. Ooi, R. Sacks-Davis: Query optimization in an extended DBMS. *Int. Conf. on Foundation of Data Organization and Algorithm*, Springer-Verlag, Lecture Notes in Computer Science 367, 247-258 (1989).

[OSM89a]   B. C. Ooi, R. Sacks-Davis, K. J. McDonell: Extending a DBMS for geographic applications. *Proc. IEEE 5th Int. Data Engineering Conf.*, L. A., 590-597 (1989).

[OSM89b]   B. C. Ooi, R. Sacks-Davis, K. J. McDonell: Spatial indexing by binary decomposition and spatial bounding (submitted for publication) (1989).

[Ora85]   Oracle overview and introduction to SQL. Oracle Corporation, U.S.A. (1985).

[Ore82]   J. A. Orenstein: Multidimensional tries for associative searching. *Information Processing Letters 14*, 4, 150-157 (1982).

[Ore86]   J. A. Orenstein: Spatial query processing in an object-oriented database system. *Proc. ACM SIGMOD International Conf. on Management of Data*, Washington, D.C., 326-336 (1986).

[OrM88]   J. A. Orenstein, F. A. Manola: PROBE spatial data modeling and query processing in an image database application. *IEEE Trans. on Software Eng. 14*, 5, 611-629 (1988).

[Oto85]   E. J. Otoo: A multidimensional digital hashing scheme for files with composite keys. *Proc. ACM SIGMOD International Conf. on Management of Data*, Austin, Texas, 214-229 (1985).

[OuS81]   M. Ouksel, P. Scheuermann: Multidimensional B-trees: Analysis of dynamic behavior. *BIT 21*, 401-418 (1981).

[Ous84]   J. K. Ousterhout: Corner stitching: A data structuring technique for VLSI layout tools. *IEEE Tans. on Comp. Aided Design CAD-3*, 1, 87-100 (1984).

[OvL82]   M. H. Overmars, J. V. Leeuwen: Dynamic multi-dimensional data structures based on quad- and KD- trees. *Acta Information 17*, 267-285 (1982).

[POL90]   H. H. Pang, B. C. Ooi, H. J. Lu: An integrated query optimizer for OODB. Tech. Rep. TR90-18-0, Institute of Systems Science, National University of Singapore (1990).

[Ric81]   P. Richard: Evaluation of the size of a query expressed in relational algebra. *Proc. ACM SIGMOD International Conf. on Management of Data*, Ann Arbor, Michigan, 155-163 (1981).

[Rob81]   J. T. Robinson: The K-D-B-tree: A search structure for large multidimensional dynamic indexes. *Proc. ACM SIGMOD International Conf. on Management of Data*, Ann Arbor, Michigan, 10-18 (1981).

[Ros85]   J. B. Rosenberg: Geographical data structures compared: A study of data structures supporting region queries, *IEEE Trans. on Comp. Aided Design CAD-4*, 1, 53-67 (1985).

[RoR82]   A. Rosenthal, D. Reiner: An architecture for query optimization. *Proc. ACM SIGMOD International Conf. on Management of Data*, Orlando, Florida, 246-255 (1982).

[RoL84]   N. Roussopoulos, D. Leifker: An introduction to PSQL: A pictorial structured query language. *IEEE Workshop on Visual Language*, Hiroshima, Japan, 77-87 (1984).

[RoL85]   N. Roussopoulos, D. Leifker: Direct spatial search on pictorial databases using packed R-trees. *Proc. ACM SIGMOD International Conf. on Management of Data*, Austin, Texas, 17-31 (1985).

[RFS88]   N. Roussopoulos, C. Faloutsos, T.K. Sellis: An efficient pictorial database system for PSQL. *IEEE Trans. on Software Eng. 14*, 5, 639-650 (1988).

[RoS79]   L. Rowe, K. Schoens: Data Abstraction, views and updates in RIGEL. *Proc. ACM SIGMOD International Conf. on Management of Data*, Boston, MA (1979).

[RoS86]   L. A. Rowe, M. Stonebraker: The commercial INGRES epilogue. In: M. Stonebraker (ed): *The INGRES Papers: Anatomy of a Relational Database System*, Reading, MA: Addison-Wesley, 63-82 (1986).

[SGR85]   R. Sacks-Davis, G.K. Gupta, K. Ramamohanarao: Dynamic hashing schemes: A review of recent developments. *Proc. 1st Pan Pacific Comp. Conf.*, Melbourne, 203-215 (1985).

[SMO87]   R. Sacks-Davis, K. J. McDonell, B. C. Ooi: GEOQL - A query language for geographic information systems. *Australian and New Zealand Association for the Advancement of Science Congress*, Townsville Australia (1987) (also appears as Tech. Rep. 87/2, Dept. of Computer Sc. R.M.I.T. Melbourne).

[Sam84]　H. Samet: The quadtree and related hierarchical data structures. *ACM Comp. Surveys 16*, 187-260 (1984).

[Sam86]　H. Samet: *Bibliography on Quadtrees and Hierarchical Data Structures.* Computer Sc. Dept., University of Maryland, Maryland (1986).

[Sam88]　H. Samet: Hierarchical representations of collections of small rectangles. *ACM Comp. Surveys 20*, 4, 271-309 (1984).

[ScO82]　P. Scheuermann, M. Ouksel: Multidimensional B-trees for associative searching in database systems. *Information Systems 7*, 2, 123-137 (1982).

[Sed77]　R. Sedgewick: Permutation generation methods. *ACM Comp. Surveys 9*, 2, 137-165 (1987).

[SeK88]　B. Seeger, H. Kriegel: Techniques for design and implementation of efficient spatial access methods. *Proc. Fourteenth Int. Conf. on Very Large Data Bases*, L.A., California, 360-371 (1988).

[SAC79]　P. G. Selinger, A. A. Astrahan, D. D. Chamberlin, R. A. Lorie, T. G. Price: Access path selection in a Relational Database Management System. *Proc. ACM SIGMOD International Conf. on Management fo Data*, Boston, MA, 23-34 (1979).

[Sel86]　T. K. Sellis: Global query optimization. *Proc. ACM SIGMOD International Conf. on Management of Data*, Washington, D.C., 191-205 (1986).

[Sel87]　T. Sellis: Private communication through email (1987).

[SRF87]　T. Sellis, N. Roussopoulos, C. Faloutsos: The R^+-tree: A dynamic index for multi-dimensional objects. *Proc. 13th Int. Conf. Very Large Data Bases*, Brighton, England, 507-518 (1987).

[Sha86]　C. A. Shaffer: Application of alternative quadtree representation. MCS-83-02118, Ph.D. Thesis, Center for Automation Research, Uni. of Maryland, College Park (1986).

[ShB78]　M. I. Shamos, J. L. Bentley: Optimal algorithm for structuring geographic data. *Proc. 1st International Advanced Study Symp. on Topological Data Structure for Geographic Inf. Sys. 6*, Dedham, Mass., Harvard Univ. Lab for Computer Graphics and Spatial Analysis (1978).

[ShH78]　L. G. Shapiro, R. M. Haralick: A general spatial structure. *Proc. Conf. on Pattern Recognition and Image Processing*, Chicago, IL, 238-249 (1978).

[Sha86]　L. D. Shapiro: Join processing in database systems with large main memories. *ACM Trans. on Database Sys. 11*, 3, 239-264 (1986).

[ShR85]   K. D. Sharma, R. Rani: Choosing optimal branching factors for k-d-b trees. *Information Systems 10*, 1, 127-134 (1985).

[SSD88]   S. Shekhar, J. Srivastava, S. Dutta: A heuristic search approach to semantic query optimization. *Proc. Int. Conf. on Very Large Data Bases* (1988).

[ShO87]   S. T. Shenoy, Z. M. Ozsoyoglu: A system for semantic query optimization. *Proc. ACM International Conf. on the Management of Data*, San Francisco, 181-195 (1987).

[SiW88]   H. Six, P. Widmayer: Spatial searching in geometric databases. *IEEE 4th International Conf. on Data Engineering*, L. A., California, 496-503 (1988).

[SmC75]   J. M. Smith, P. Y. Chang: Optimizing the performance of a relational algebra database interface. *Comm. ACM 18*, 568-579 (1975).

[Sor84]   J. J. Sordi: The query management facility. *IBM Sys. J. 23*, 2, 126-150 (1984).

[SWK76]   M. Stonebraker, E. Wong, P. Kreps, G. Held: The design and implementation of INGRES. *ACM Trans. on Database Sys. 1*, 3, 189-222 (1976).

[SRG83]   M. Stonebraker, B. Rubenstein, A. Guttman: Applications of abstract data types and abstract indices to CAD data. *Proc. Engineering Design Applications of ACM-IEEE Database Week*, San Jose, California, 107-115 (1983).

[StG84]   M. Stonebraker, A. Guttman: Using a relational database management system for computer aided design data - An update. *IEEE Database Engineering 7*, 2, 56-60 (1984).

[SAH84]   M. Stonebraker, M. Anderson, E. Hanson, N. Rubenstein: QUEL as a data type. *Proc. ACM SIGMOD International Conf. on Management of Data*, Boston, MA, 208-214 (1984).

[Sto86]   M. Stonebraker: Inclusion of new types in relational data base systems. *Proc. 2nd International Conf. on Data Base Engineering*, LA, California, 262-269 (1986).

[StR86]   M. Stonebraker, L. A. Rowe: The design of POSTGRES. *Proc. ACM SIGMOD International Conf. on the Management of Data*, Washington, D.C., 340-355 (1986).

[StR87]   M. Stonebraker, L. Rowe: The POSTGRES data model. *Proc. 13th Int. Conf. on Very Large Data Bases*, Brighton, England, 83-96 (1987).

[TII80]   Y. Takao, S.Itoh, J. Iisaka: An image-oriented data base system. In: *Data Base Techniques for Pictorial Applications*, Springer-Verlag, 527-538 (1980).

[Tam82a]   M. Tamminen: Efficient spatial access to a data base. *Proc. SIGMOD International Conf. on the Management for Data*, ACM, 200-206 (1982).

[Tam82b]   M. Tamminen: The extendible cell method for closest points problems. *BIT 22*, 27-41 (1982).

[Tam83]   M. Tamminen: Performance analysis of cell based geometric file organisations. *International Journal Comp. Vision, Graphics, and Image Processing 24*, 160-181 (1983).

[TaS82]   M. Tamminen, R. Sulonen: The EXCELL method for efficient geometric access to data. *Proc. ACM IEEE 19th Design Automation Conf.*, Las Vegas, Nevada, 345-351 (1982).

[TKN80]   T. Tsurutani, Y. Kasahara, M. Naniwada: ATLAS: A geographic database system - data structure and language design for geographic information. *Comp. Graphics*, ACM, New York, 71-77 (1980).

[Val86]   P. Valduriez: Optimization of complex database queries using join indices. In: G. Lohman (ed): *A quarterly bulletin of the Computer Society of the IEEE technical committee on Database Engineering 9*, 4, 10-16 (1986).

[Val87]   P. Valduriez: Join indices. *ACM Trans. on Database Sys. 12*, 2, 218-246 (1987).

[WhK85]   K. Whang, R. Krishnamurthy: Multilevel grid files. Report RC 11516, IBM Thomas J. Watson Research Center, Yorktown Heights, New York (1985).

[WoY76]   E. Wong, K. Youssefi: Decomposition - A strategy for query processing. *ACM Trans. on Database Sys. 1*, 3, 223-241 (1976).

[Yao79]   S. B. Yao: Optimization of query evaluation algorithms. *ACM Trans. on Database Sys. 4*, 2, 133-155 (1979).

[YoW79]   K. Youssefi, E. Wong: Query processing in a relational database management system. *Proc. 5th Int. Conf. on Very Large Data Bases*, Rio de Janeiro, Brazil, 409-417 (1979).

[YuO79]   C. T. Yu, M. E. Ozsoyoglu: An algorithm for tree query membership of a distributed query. *Proc. IEEE COMPSAC - Comp. Software & Applications Conf.*, Chicago, IL, 306-312 (1979).

[Zan83]   C. Zaniolo: The database language GEM. *Proc. ACM SIGMOD International Conf. on Management of Data*, San Jose, California, 207-218 (1983).

[Zip49]   G. K. Zipf: *Human behaviour and the principle of least effort*. Addison-Wesley, Cambridge, MA (1949).

[Zlo76]   M. M. Zloof: Query-by-Example: Operations on hierarchical data bases. *Proc. National Comp. Conf. 45*, 845-853 (1976).

Appendix A

GEOQL Grammar

The following grammar for GEOQL uses a BNF-style notation in which keywords appear in **UPPERCASE BOLD** and non-terminals symbols are in *lowercase italic*. Several special lexicons are also used, namely

identifier	A table name, correlation name or field name.
number	A numeric constant.
string	A constant, enclosed in quotes (").

The symbol ':' introduces the right-hand side of a production rule, and '|' designates an alternative production. All other non-alphanumeric characters are literal symbols.

Operator Precedence		
Precedence	Operator(s)	Associativity
1 (low)	**GROUP HAVING WHERE**	left
2	**THEN**	left
3	**OR**	left
4	**AND**	left
5	**NOT**	right
6	'+' '-'	left
7	'/'	left
8	'*'	left
9 (high)	unary '+' '-'	right

geoql_statement   :   *cursor_spec* /

cursor_spec   :   *query_expr order_by_clause*

query_expr   :   *query_term*
        | *query_expr* **UNION** *all_opt query_term*

query_term   :   *query_spec*
        | (*query_expr*)

query_spec	:	**SELECT** *all_distinct select_list table_expr*
all_distinct	:	
	|	**ALL**
	|	**DISTINCT**
select_list	:	*sel_expr_list*
	|	*
sel_expr_list	:	*value_expr*
	|	*sel_expr_list* , *value_expr*
value_expr	:	*term*
	|	*value_expr addop term*
addop	:	+
	|	–
term	:	*factor*
	|	*term mulop factor*
mulop	:	/
	|	*
factor	:	*primary*
	|	*addop primary*
primary	:	*value_spec*
	|	*column_spec*
	|	*set_func_spec*
	|	(*value_spec*)
value_spec	:	*literal*
literal	:	string
	|	number
column_spec	:	*column_name*
	|	*qualifier* . *column_name*
	|	*column_name* . *column_name*
column_name	:	identifier
qualifier	:	*table_name*

table_name	:	identifier
set_func_spec	:	**COUNT (*)**
	|	**COUNT (DISTINCT** *column_spec*)
	|	*distinct_set_func*
	|	*all_set_func*
distinct_set_func	:	*s_f* (**DISTINCT** *column_spec*)
s_f	:	**AVG**
	|	**MAX**
	|	**MIN**
	|	**SUM**
all_set_func	:	*s_f* (*all_opt value_expr*)
table_expr	:	*from_clause where_clause group_by_clause having_clause*
from_clause	:	**FROM** *from_list*
from_list	:	*f_l_ent*
	|	*from_list* , *f_l_ent*
f_l_ent	:	*table_name correl_name_opt*
correl_name_opt	:	
	|	*correlation_name*
correlation_name	:	identifier
where_clause	:	
	|	**WHERE** *search_condition*
search_condition	:	*boolean_term*
	|	*search_condition* **OR** *boolean_term*
boolean_term	:	*boolean_factor*
	|	*boolean_term* **AND** *boolean_factor*
boolean_factor	:	**NOT** *boolean_primary*
	|	*boolean_primary*
boolean_primary	:	*predicate*
	|	(*search_condition*)

predicate	:	*comparison_pred*
	|	*between_pred*
	|	*in_pred*
	|	*null_pred*
	|	*exists_pred*
	|	*geo_pred*
comparison_pred	:	*nucleotide compop nucleotide*
nucleotide	:	*value_expr*
	|	*sub_query*
	|	*quant sub_query*
sub_query	:	(**SELECT** *all_distinct result_spec table_expr*)
result_spec	:	*value_expr*
	|	*
quant	:	**ALL**
	|	**SOME**
	|	**ANY**
compop	:	!=
	|	<
	|	>
	|	<=
	|	>=
	|	=
between_pred	:	*value_expr not_opt* **BETWEEN** *value_expr* **AND** *value_expr*
not_opt	:	
	|	**NOT**
in_pred	:	*value_expr not_opt* **IN** *sub_query*
	|	*value_expr not_opt* **IN** (*in_value_list*)
in_value_list	:	*value_spec*
	|	*in_value_list* , *value_spec*
null_pred	:	*column_spec* **IS** *not_opt* **NIL**
exists_pred	:	**EXISTS** *sub_query*

geo_pred	:	*geo_obj* **non_window_pred** *geo_term*
	|	*window_term* **window_pred** *geo_term*
	|	*geo_term* **distance_pred** *window_term*
non_window_pred	:	**ENDS** *at_opt*
	|	**JOINS**
	|	*is_opt* **ADJACENT** *to_opt*
	|	*is_opt* **SITUATED** *at_opt*
	|	*is_opt* **WITHIN** number *of_opt*
window_pred	:	**INTERSECTS**
	|	**CONTAINS**
distance_pred	:	*is_opt* **CLOSEST** *to_opt*
	|	*is_opt* **FURTHEST** *from_opt*
window_term	:	*geo_term*
	|	**WINDOW**
	|	*window*
geo_term	:	*geo_obj*
	|	**LINE JOINING** *geo_obj* **AND** *geo_obj*
	|	*geo_obj* **BOUNDED BY** *geo_obj* **AND** *geo_obj*
geo_obj	:	*qualifier*
window	:	*(value_expr, value_expr, value_expr, value_expr)*
at_opt	:	
	|	**AT**
is_opt	:	
	|	**IS**
to_opt	:	
	|	**TO**
of_opt	:	
	|	**OF**
from_opt	:	
	|	**FROM**

group_by_clause   :
                      | **GROUP BY** *column_list*

column_list   :   *column_spec*
                      | *column_list , column_spec*

having_clause   :
                      | **HAVING** *search_condition*

order_by_clause   :
                      | **ORDER BY** *sort_spec*

sort_spec   :   *column_spec direction*
                      | *column_spec direction , sort_spec*

direction   :
                      | **ASC**
                      | **DESC**

all_opt   :
                      | **ALL**

Appendix B

Query Optimization Algorithms

In the implementation of a multiway tree, the left-child-right-sibling method is used. However, for the description of the decomposition algorithm, a node of the strategy tree is assumed to have an array of child nodes. Note that the grouping of predicates and checking of whether a join is an existential restriction is done in the parsing phase.

Algorithm Decompose

Input:	p_node — an intermediate predicate tree node.
	s_node — An intermediate strategy tree node.
Output:	A strategy tree.
Comment:	The algorithm decomposes a predicate tree into several subtrees, and for each subtree the subquery is created to represent that subtree. All subqueries appear as leaf nodes in a strategy tree.

DECOMPOSE(s_node, p_node)

let the expression of the subtree of p_node be P and let $P = \log\_op(P_1, ..., P_k)$ where each P_i ($1 \leq i \leq k$) is the expression of a child of p_node.

count =1;

if there does not exist any P^s in P **then**

   return;

for each P_i^s **do**

   FORMSUB(P_i^s, s_node, count, US);

if $\log\_op = and$ **then**

   /* K = a set of relations involved in spatial predicates of lower levels */

   $$K = \bigcup_{P_u^s \text{ in the subtree of } P} S_u;$$

   /* Subqueries of restrictions */

   for each restriction P_i, ($|S_i| = 1$) and ($S_i \cap K$) **do**

      if there is any existential restriction whose $S \cap S_i$ **then**

let the restrictions be P_j^a, ..., P_l^a;

  FORMSUB(log_op(P_i^a, P_j^a, ... P_l^a)), s_node, count, UA);

else

  FORMSUB(P_i, s_node, count, UA);

 /* *Subqueries of joins.* */

 for each P_i^a given that $|S_i| \neq 1$ and $S_i \cap K$ **do**

    FORMSUB(P_i^a, s_node, count, UA);

Let all the remaining P_i^a of P be P_m^a, ..., P_n^a;

FORMSUB(log_op(P_m^a, ..., P_n^a), s_node, count, UA);

for each remaining P_i **do**

 /* *Now, only compound predicates* of the form log_op(P_{i1}, ..., P_{ij})

   that contain spatial predicates remain */

   new_s_node = a new strategy tree node;

   new_s_node.type = log_op of P_i;

   DECOMPOSE(new_s_node, root of P_i).

   if new_s_node.child[1] $\neq$ NULL **then**

    /* *either NULL or at least two children* */

      s_node.child[count++] = new_s_node;

end DECOMPOSE.

FORMSUB(P, s_node, count, query_type)

Input: P — a subtree of predicates where P is the root node.

          s_node — a strategy tree node.

          count — the current position of the children of the s_node.

          query_type — the query type (US or UA).

Output: updated s_node and count.

Comment: Given a subtree of predicates of the WHERE clause, the
          FORMSUB forms a query tree for that subtree and makes the query
          as a child of the strategy tree node s_node.

form a GEOQL query Q with

   P as the qualification (the WHERE clause);

   The relations in P be the referenced relations in the FROM clause;

   A unique temporary relation name in the INTO clause;

s_node.child[count].type = query_type;

s_node.child[count++].query = Q;

end FORMSUB.

Subquery sequences are generated using the following the permutator which generates the sequences in lexicographically order. We follow the method advocated byEr87.

LEXPERM(i, p, q, N)

Input: i - initially 1;

p - the array that stores the current sequence.

q - the array that stores the elements.

N - the number of elements;

Output: A sequence of numbers indicating the SQP.

/* *The permutator which generates numbers in lexicographically order.* */

if i = N-1 **then**

        p[i] = s[0];

        p[N] = s[s[0]];

        EXAMINE(p, q, N);

        p[i]=s[s[0]];

        p[N]=s[0];

        EXAMINE(p, q, N);

else

        y=0;

        x= s[0];

        while x ≤ N **do**

                x=s[y];

                p[i]=x;

                s[y]=s[x];

                LEXPERM(i+1, p, s, N);

                s[y]=x;

                y=x;

                x=s[x];

end LEXPERM.

The Formulate algorithm makes use of the LEXPERM routine to generate different SQPs.

Algorithm Formulate

FORMULATE(snode, cur_pos)

Input: snode - a logical node the strategy tree; initially the root.

              cur_pos - the current position of SQP array; initially 0.

Output: A near optimal subquery sequence.

Parameters:

              K - SQP array.

              p - array describes the subsequence of subqueries of snode.

              s - working array;

              $cost_{best}$ - the best cost obtained so far; initially a huge number.

              cost - cost of the current sequence; initially 0.

              reject - the number of subquery sequences generated subsequently

              that will be rejected

N_{nodes} = number of child nodes of snode;

Let the nodes be $node_i$ ($1 \le i \le N_{nodes}$);

for i=1 to N_{nodes} **do**

    p[i] = i;

    s[i] = i+1;

    q[i] = $node_i$;

reject = 0;

if snode is a logical *and* node **then**

    LEXPERM(1, p, q, s, N_{nodes}, cur_pos);

else

    /* *An or node.* */

    EXAMINE(p, q, N_{nodes}, cur_pos);

end FORMULATE.

EXAMINE(p, q, N, cur_pos)

Input: p - integer array describes a sequence of subqueries.

q - array describe a random order of a set of strategy tree nodes.

N - the number of strategy tree nodes in q.

cur_pos - the current position in the current SQP array.

Output: The best SQP obtained so far.

Parameter: Tcost - temporary storage used to record the cost of subsequence

obtained so far before more subqueries are added to it.

if reject $\neq$ 0 **then**

reject = reject-1;   /* *reject the sequence* */

return;

final = true; Tcost = cost;

for i = 1 to N **do**

K[i--cur_pos] = q[p[i]];

for i=1 to N **do**

if q[p[i]] is a logical node **then**

FORMULATE(q[p[i]], cur_pos+N);

final = false;

else /* *a query node* */

select referencing relations that incur least cost;

attributes that are required for further evaluation and the answer

are selected in the SELECT clause;

cost = cost + Cost($Q_{p[i]}$);

check it against the heuristic rules 1 and 3;

check it for correctness, following result rewriting rule 1 - 3;

/* *Reject all sequences with the same initial subsequence* */

if cost > cost$_{best}$ or rejected because of the heuristic rules **then**

reject = (N - i)! - 1;

/* *Incremental selection.* */

if (final) **then**

/* *When final is true, the sequence is a complete one.* */

if (cost$_{best}$ < cost) **then**

cost$_{best}$ = cost;

save the sequence;

cost = Tcost;

end EXAMINE.

Appendix C

Sample Database

Lake

gid	name	a2
0011	Eildon	11
0012	West	11
0013	Silvan	21
0014	Albert	22
0015	Lara	34
0016	Sea	35

Lakespatial

gid	mbr
0011	12:20:50:62
0012	8:16:55:70
0013	10:15:99:110
0014	2:5:50:60
0015	70:79:6:12
0016	20:24:82:88

Region

gid	name	a1
0001	region1	11
0002	region2	11
0003	region3	21
0004	region4	22
0005	region5	34
0006	region6	35
0007	region7	75
0008	region8	58
0009	monash	66
0010	rmit	67
0011	csiro	45
0012	telecom	91
0013	meluni	54
0014	cit	18
0015	clayton	67
0016	carlton	35
0017	fitzroy	75
0018	ashwood	56
0019	rusden	42
0020	boxhill	87

Regionspatial

gid	mbr
0001	10:20:50:60
0002	8:18:50:70
0003	0:5:99:110
0004	22:35:50:60
0005	90:99:6:12
0006	40:44:80:88
0007	20:25:45:60
0008	100:110:50:80
0009	15:56:180:189
0010	20:36:68:74
0011	88:102:156:167
0012	67:89:155:178
0013	34:56:20:40
0014	108:120:40:67
0015	178:182:134:144
0016	16:54:167:178
0017	56:78:78:88
0018	126:134:167:190
0019	17:28:48:92
0020	47:58:84:92

Railway

gid	name	a4
1001	railway1	01
1002	railway2	11
1003	railway3	01
1004	railway4	22
1005	railway5	04
1006	railway6	85

Railwayspatial

gid	mbr
1001	15:25:55:65
1002	8:18:50:70
1003	10:25:99:110
1004	22:35:50:60
1005	90:99:6:12
1006	80:84:90:98

Road

gid	name	a3
1011	road1	11
1012	road2	11
1013	road3	21
1014	road4	22
1015	road5	34
1016	road6	35
1017	road7	78
1018	road8	86
1019	princess hwy	78
1020	high st	54
1021	blackburn rd	75
1022	queen st	45
1023	russel st	43
1024	kooyong rd	18
1025	brown rd	34

Roadspatial

gid	mbr
1011	12:20:50:62
1012	8:16:55:70
1013	10:15:99:110
1014	2:5:50:60
1015	70:79:6:12
1016	20:24:82:88
1017	20:78:10:16
1018	67:88:99:112
1019	10:101:89:96
1020	56:89:56:93
1021	102:150:120:146
1022	112:178:187:199
1023	16:67:36:78
1024	17:18:25:45
1025	46:57:77:89

This series reports new developments in computer science research and teaching – quickly, informally and at a high level. The type of material considered for publication includes preliminary drafts of original papers and monographs, technical reports of high quality and broad interest, advanced level lectures, reports of meetings, provided they are of exceptional interest and focused on a single topic. The timeliness of a manuscript is more important than its form which may be unfinished or tentative. If possible, a subject index should be included. Publication of Lecture Notes is intended as a service to the international computer science community, in that a commercial publisher, Springer-Verlag, can offer a wide distribution of documents which would otherwise have a restricted readership. Once published and copyrighted, they can be documented in the scientific literature.

Manuscripts

Manuscripts should be no less than 100 and preferably no more than 500 pages in length.

They are reproduced by a photographic process and therefore must be typed with extreme care. Symbols not on the typewriter should be inserted by hand in indelible black ink. Corrections to the typescript should be made by pasting in the new text or painting out errors with white correction fluid. Authors receive 75 free copies and are free to use the material in other publications. The typescript is reduced slightly in size during reproduction; best results will not be obtained unless the text on any one page is kept within the overall limit of 18 x 26.5 cm (7 x 10½ inches). On request, the publisher will supply special paper with the typing area outlined.

Manuscripts should be sent to Prof. G. Goos, GMD Forschungsstelle an der Universität Karlsruhe, Haid- und Neu-Str. 7, 7500 Karlsruhe 1, Germany, Prof. J. Hartmanis, Cornell University, Dept. of Computer Science, Ithaca, NY/USA 14850, or directly to Springer-Verlag Heidelberg.

Springer-Verlag, Heidelberger Platz 3, D-1000 Berlin 33
Springer-Verlag, Tiergartenstraße 17, D-6900 Heidelberg 1
Springer-Verlag, 175 Fifth Avenue, New York, NY 10010/USA
Springer-Verlag, 37-3, Hongo 3-chome, Bunkyo-ku, Tokyo 113, Japan

ISBN 3-540-53474-1
ISBN 0-387-53474-1